NEUROSCIENCE, NEUROPSYCHOLOGY, NEUROPSYCHIATRY, BRAIN & MIND:
Primer, Overview & Introduction

by R. Joseph, Ph.D.

NEUROSCIENCE, NEUROPSYCHOLOGY, NEUROPSYCHIATRY, BRAIN & MIND:
Primer, Overview & Introduction

by R. Joseph, Ph.D.

University Press Science Publishers

Copyright © 2000, 2011, 2012
R. Joseph, Ph.D.

Published by: University Press Science Publishers

ISBN: 9780970073334

Contents

NEUROSCIENCE, NEUROPSYCHOLOGY, NEUROPSYCHIATRY, BRAIN & MIND: Primer, Overview & Introduction   16

Overview of the Brain and Mind: Functional Localization   16
Localization & Functional Neuroanatomy Of The Brain   21
The Old Cortex & the New Cortex   22
Primary Receiving Areas   26
Frontal Lobe Monitoring of Activity 29
Knowing Yet Not Knowing: Disconnected Consciousness 30
The Visual Mind: Denial of Blindness 32
"Blind Sight" 35
Body Consciousness: Denial of the Body, and Phantom Limbs   37
Summary: Brain & Mind.   41
The Brainstem 43
Brainstem Cranial Nerves   49
CRANIAL NERVE 12: HYPOGLOSSAL 50
CRANIAL NERVE 11: SPINAL ACCESSORY   51
CRANIAL NERVE 10: VAGUS 51
CRANIAL NERVE 9: GLOSSOPHARYNGEAL   51
THE EIGHT CRANIAL NERVE & THE COCHLEAR SYSTEM   52
TINNITUS, DEAFNESS, DIZZINESS, VERTIGO 52
THE 8th (VESTIBULAR) NERVE   54
ORAL-FACIAL MOVEMENT & SENSATION   55
THE 7th CRANIAL (FACIAL) NERVE   55
THE 5th CRANIAL NERVE: TRIGEMINAL   55
EYE MOVEMENT: THE 6th CRANIAL NERVE: ABDUCENS   56
CRANIAL NERVES OF THE MIDBRAIN EYE MOVEMENT   57
CRANIAL NERVE 4: TROCHLEAR   57
CRANIAL NERVE 3: OCULOMOTOR   57
THE OPTIC NERVE   58
THE OLFACTORY NERVE   59
The Cerebellum   61
The Diencaphalon: Hypothalamus, Thalamus   66
The Limbic System: Emotion & Motivation 70
Amygdala, Hippocampus, Septal Nuclei, Cingulate, Hypothalamus.   70
Limbic System Sexuality   72
Social Behavior & The Limbic System   75
Amygdala, Hippocampus & Memory   75

Long Term Potentiation & Memory   77
Hippocampus, Memory & Amnesia   80
The Cingulate & Entorhinal Cortex   81
The Limbic And Corpus Striatum   83
The Limbic System Vs Neocortex: Consciousness   87
The Neocortex (Gray Matter) 88
Neocortical Layers   90
Cytoarchitextural, Neuronal, & Chemical Organization of the Neocortex   96
Neocortex & The Conscious And Unconscious (Emotional) Mind   103
The Frontal, Parietal, Temporal And Occipital Lobes105
The Frontal Lobes   106
The Frontal Motor Areas   111
The Temporal Lobes   115
The Parietal Lobes   121
The Occipital Lobes And Vision   127
The Primary, Secondary And Association Areas   132
The Split-Brain. Right And Left Hemisphere: Functional Laterality 137
Dissociation and Self-Consciousness   141
The Neuroanatomy Of Mind 144
Overview: Consciousness, Awareness, And The Neuroscience Of Mind   145

REFERENCES   147

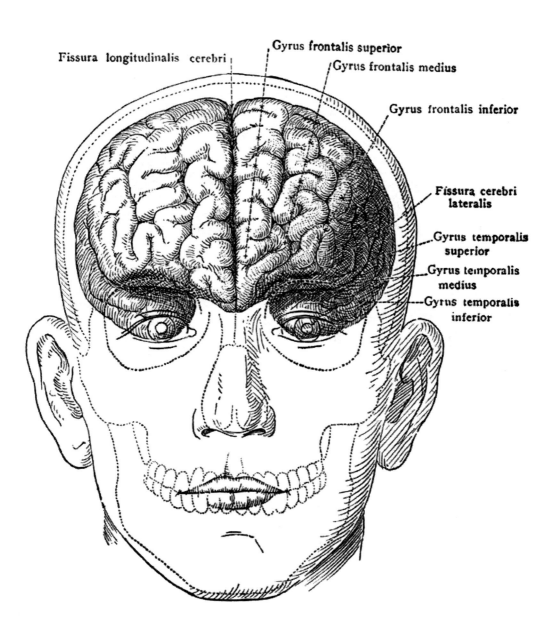

Fissura longitudinalis cerebri

Gyrus frontalis superior

Gyrus frontalis medius

Gyrus frontalis inferior

Fissura cerebri lateralis

Gyrus temporalis superior

Gyrus temporalis medius

Gyrus temporalis inferior

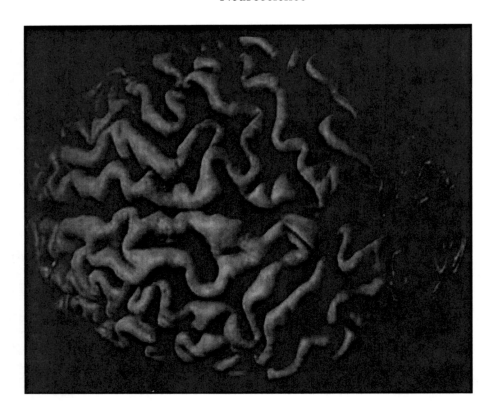

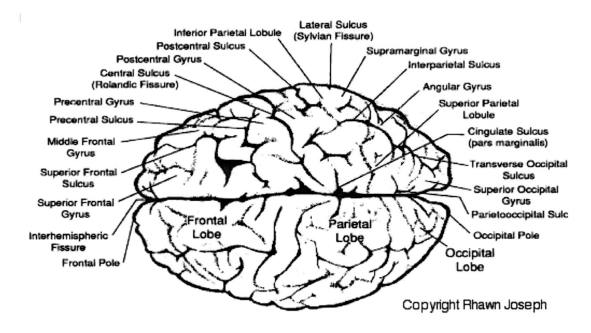

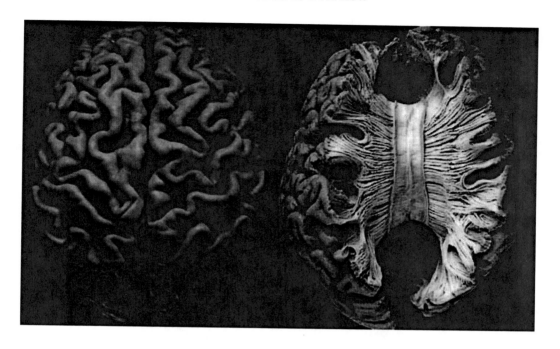

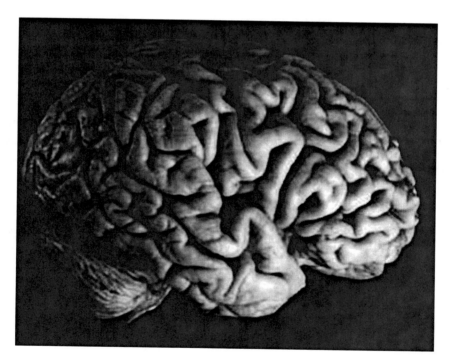

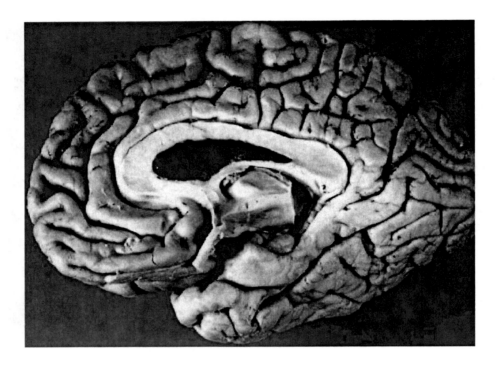

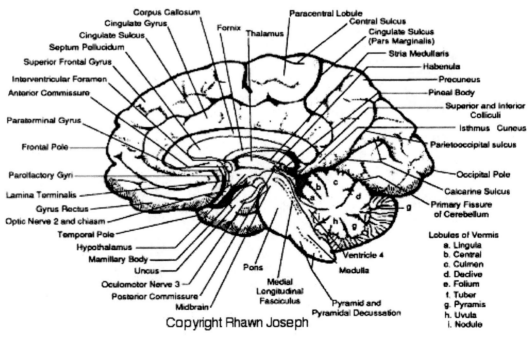

Corpus Callosum
Cingulate Gyrus
Cingulate Sulcus
Septum Pellucidum
Superior Frontal Gyrus
Interventricular Foramen
Anterior Commissure
Paraterminal Gyrus
Frontal Pole
Parolfactory Gyri
Lamina Terminalis
Gyrus Rectus
Optic Nerve 2 and chiasm
Temporal Pole
Hypothalamus
Mamillary Body
Uncus
Oculomotor Nerve 3
Posterior Commissure
Midbrain

Fornix   Thalamus

Paracentral Lobule
Central Sulcus
Cingulate Sulcus
(Pars Marginalis)
Stria Medullaris
Habenula
Precuneus
Pineal Body
Superior and Inferior Colliculi
Isthmus   Cuneus
Parietooccipital sulcus
Occipital Pole
Calcarine Sulcus
Primary Fissure of Cerebellum

Ventricle 4
Medulla

Pons
Medial Longitudinal Fasciculus
Pyramid and Pyramidal Decussation

Lobules of Vermis
a. Lingula
b. Central
c. Culmen
d. Declive
e. Folium
f. Tuber
g. Pyramis
h. Uvula
i. Nodule

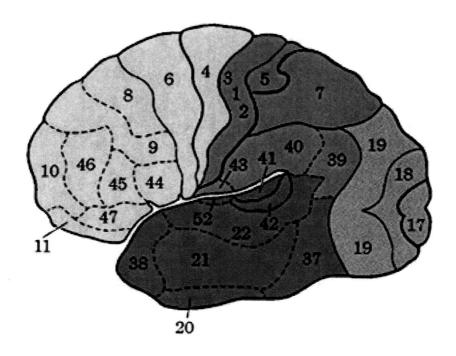

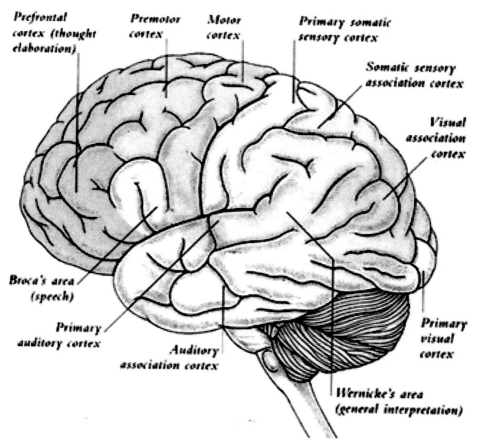

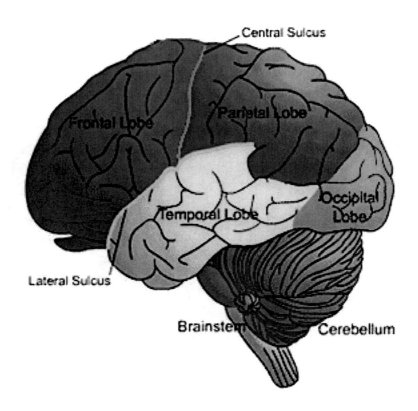

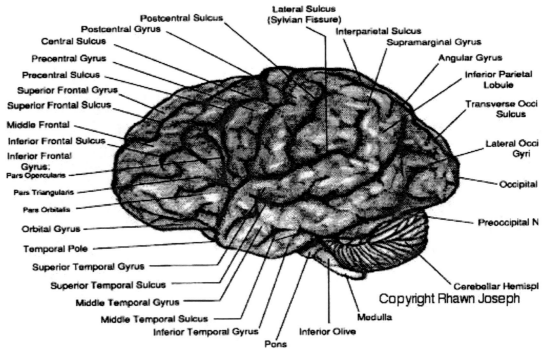

Copyright Rhawn Joseph

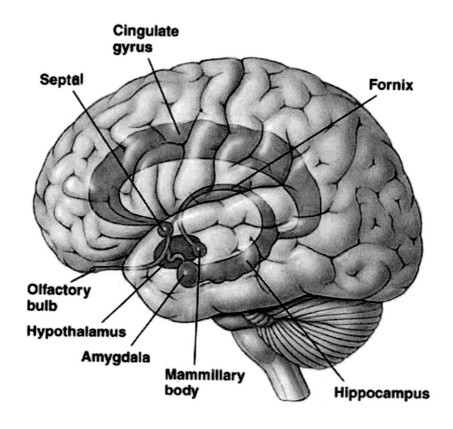

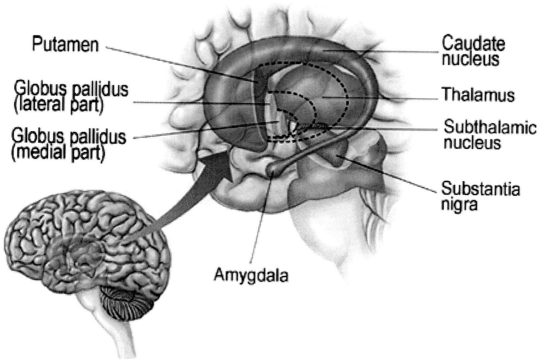

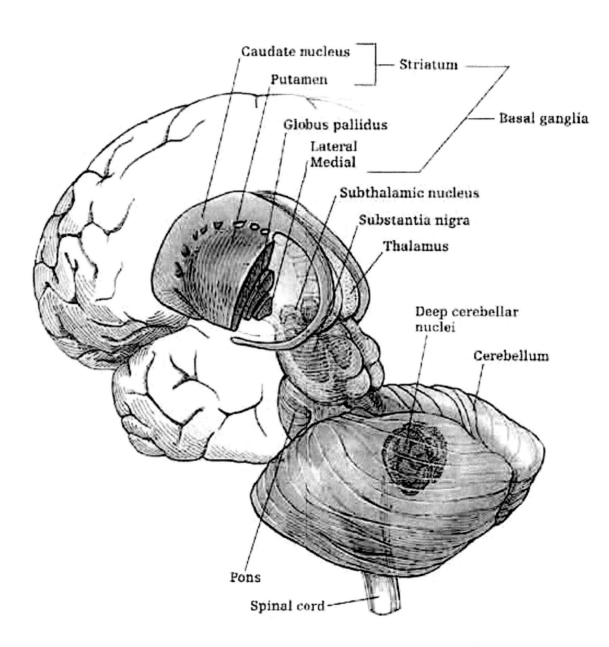

Caudate nucleus

Striatum

Putamen

Basal ganglia

Globus pallidus

Lateral
Medial

Subthalamic nucleus

Substantia nigra

Thalamus

Deep cerebellar
nuclei

Cerebellum

Pons

Spinal cord

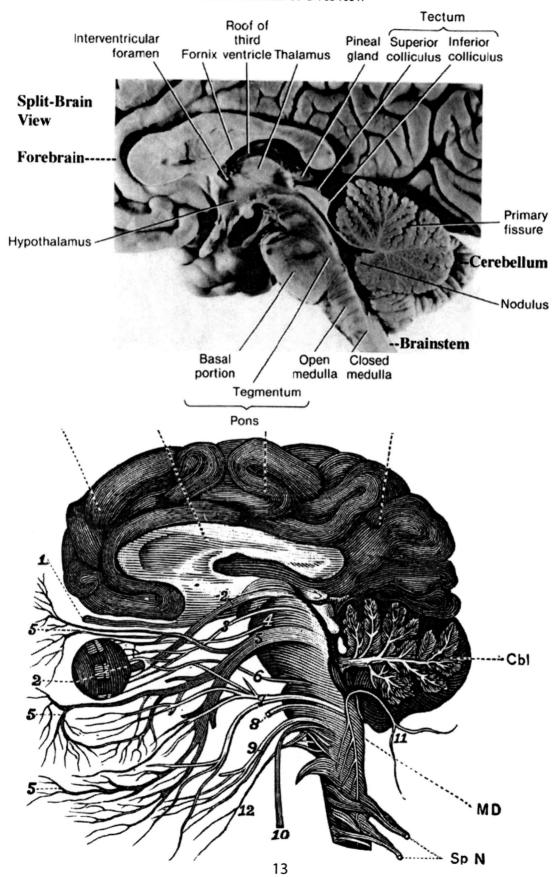

Split-Brain View

Forebrain-----

Interventricular foramen

Roof of third ventricle

Fornix

Thalamus

Pineal gland

Tectum
Superior colliculus
Inferior colliculus

Primary fissure

Cerebellum

Nodulus

Hypothalamus

Basal portion

Tegmentum

Open medulla

Closed medulla

Brainstem

Pons

Cbl

MD

Sp N

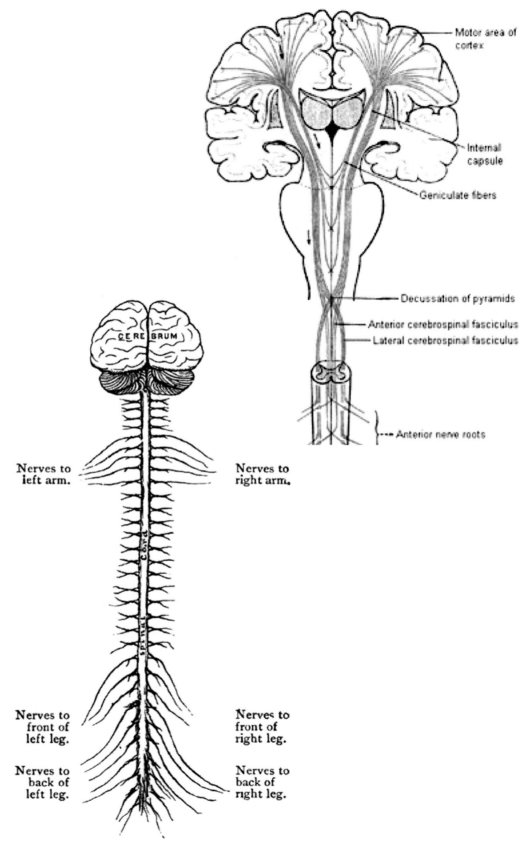

Motor area of
cortex

Internal
capsule

Geniculate fibers

Decussation of pyramids

Anterior cerebrospinal fasciculus

Lateral cerebrospinal fasciculus

Anterior nerve roots

CEREBRUM

Nerves to
left arm.

Nerves to
right arm.

Nerves to
front of
left leg.

Nerves to
front of
right leg.

Nerves to
back of
left leg.

Nerves to
back of
right leg.

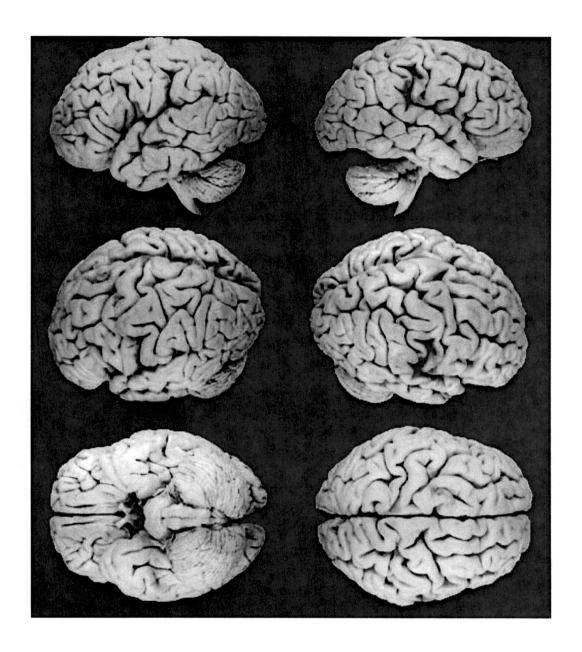

NEUROSCIENCE, NEUROPSYCHOLOGY, NEUROPSYCHIATRY, BRAIN & MIND

Primer, Overview & Introduction

Overview of the Brain and Mind: Functional Localization

The functional neuroanatomy of the human brain is functionally localized, such that specific areas of the brain subserve specific functions. For example, if you were to suffer a severe injury to the left frontal-temporal region of your head and brain, you might lose the ability to talk. This is because, the left frontal tissues known as Broca's Area mediates the expression of speech. However, you would still be able to swear and sing because these abilities are mediated by the right frontal-temporal region of the brain

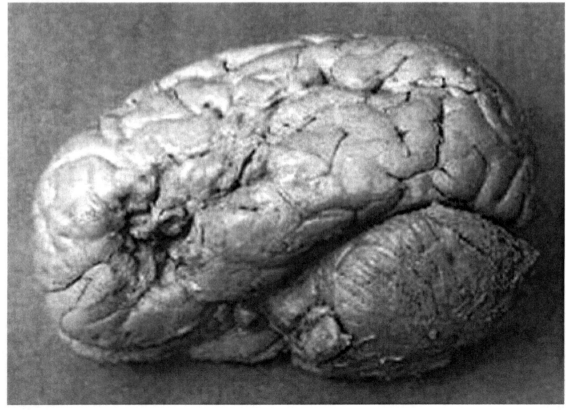

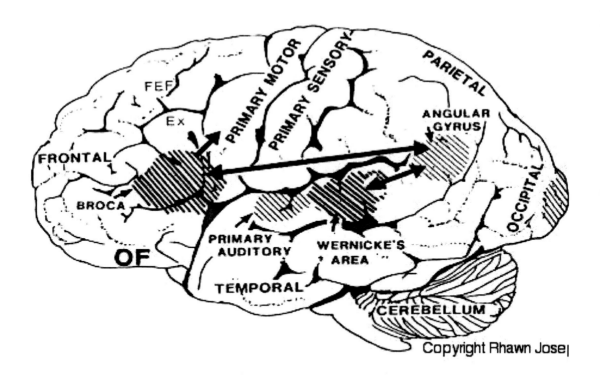

If instead, you suffered a massive stroke to the left temporal lobe (the brain area inside your skull near your ear), you would be unable to comprehend human speech. You would not understand what is being said to you. This is because the left temporal region, known as Wernicke's Area, subserves the ability to understand language.

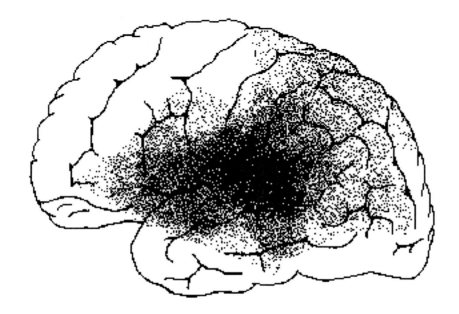

However, Wernicke's area only comprehends the words and temporal sequences of spoken language, not the melody. It is the right temporal lobe which perceives the melody and emotional qualities of the speaker's voice. Therefore, if someone were to approach you in a nightclub or bar, and say: "Do you want to step outside?" it is the left temporal lobe which understand a question concern "outside" is being asked, but it is the right temporal lobe which understand the emotion and context, what the speaker means. Thus you know if you are being asked outside to smoke a cigarette, to talk privately, or if you are being threatened with a punch in the nose.

It is the right temporal lobe which perceives emotion, melody, music, and sounds from the environment. Thus, if the right temporal lobe were damaged, the ability to hear music, sarcasm, emotion, or sounds from the environment would be lost to varying degrees depending on the extent of the injury. For example, if birds were singing, water was dripping from a faucet, the telephone were ringing, or even if someone were knocking on the door--these sounds may no longer be perceived with extensive right temporal lobe damage.

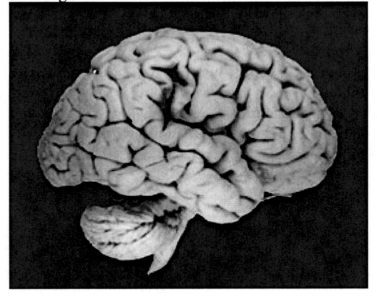

And just as the right temporal lobe perceives the melody and environmental sounds, it is the right frontal ares which sing and provides the prosodic, emotional qualities to the voice. Therefore, if the right frontal lobe were injured, the voice may lose its emotional qualities and might sound blunted and dulled. Certainly, those with extensive right frontal lobe damage may no longer be able to sing.

These brain areas do not act in isolation, however. The mind is a

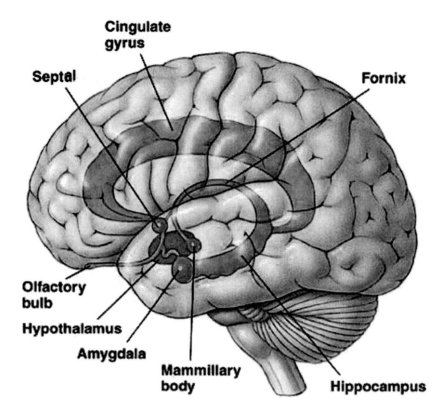

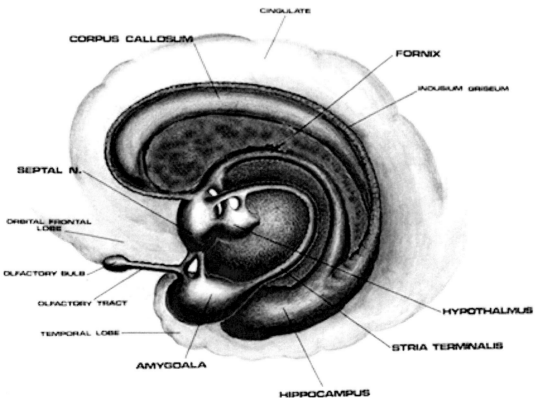

multiplicity, and different regions act together. It takes both halves of the brain to make music, the left frontal providing the words and temporal sequences, the right temporal perceiving the music, the right frontal providing the melody, and the left temporal perceive the words and rhythm.

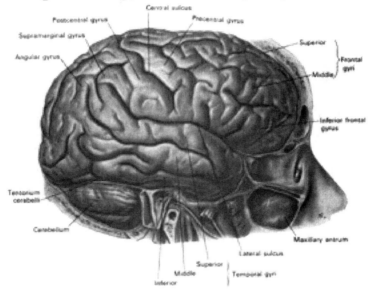

Broadly considered, there are two frontal, temporal, parietal, and occipital lobes, one in each hemisphere of the brain. Thus, the brain is a duality, as well as a multiplicity, as there are two hemispheres, a right and a left.

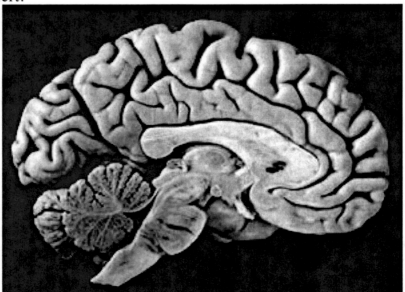

Split-brain view, showing the medial regions of the left hemisphere.

Localization & Functional Neuroanatomy Of The Brain

The human brain can be broadly described as consisting of 1) two cerebral hemispheres which are divided into frontal, parietal, temporal, and occipital lobes, 2) "gray matter" which covers the surface of the brain, and which consists of 6-7 major layers and several sublayers of specialized neurons, 3) subcortical structures consisting of 2-5 layers, and which include the striatum (caudate, putaman, globus pallidus), limbic system (hypothalamus, amygdala, hippocampus, septal nuclei, cingulate gyrus), thalamus, cerebellum, and brainstem (midbrain, pons, medulla).

The human brain is thus a multiplicity with specific areas serving specific functions. However, even specific regions serve multiple functions. For example the frontal lobes, located at the front of the brain, not only speaks and sings, but maintains attention and concentration and acts to filter out and suppress sensory and perceptual information. In fact, much of the brain works by inhibition, otherwise, the mind would be overwhelmed by sensory impressions. Consider, by way of example, you are sitting in your office reading this text. The pressure of the chair, the physical sensations of your shoes and clothes, the musculature of your body as it holds one then another position, the temperature of the room, various odors and fragrances, a multitude of sounds, visual sensations from outside your area of concentration and focus, and so on, are all being transmitted to the brainstem, midbrain, and olfactory limbic system. These signals are then relayed to various subnuclei within the thalamus.

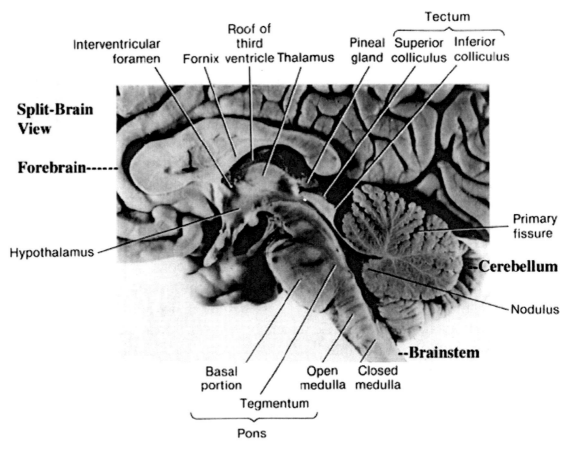

The Old Cortex & the New Cortex

The neural tissues of the brainstem, midbrain, limbic system and thalamus are associated with the "old brain." However, those aspects of consciousness we most closely associated with humans are associated with the "new brain" the neocortex (Joseph, 1982, 1992). Therefore, although you may be "aware" of these sensations while they are maintained within the old brain, you are not "conscious" of them, unless a decision is made to become conscious or they increase sufficiently in intensity that they are transferred to the neocortex via the the thalamus and frontal lobes, and forced into the focus of consciousness (Joseph, 1982, 1986b, 1992, 1999a, 2009).

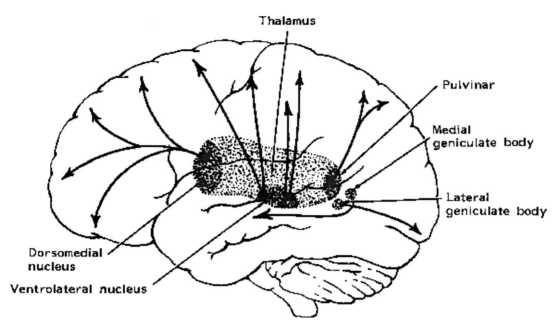

The old brain is covered by a gray mantle of new cortex, neocortex. The sensations alluded to are transferred from the old brain to the thalamus which relays these signals to the neocortex. Human consciousness and the "higher" level of the multiplicity of mind, are associated with the "new brain."

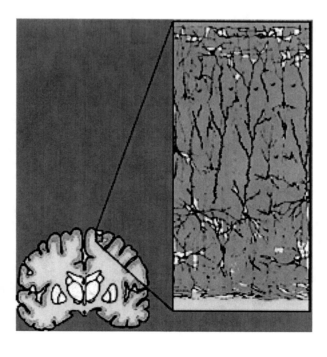

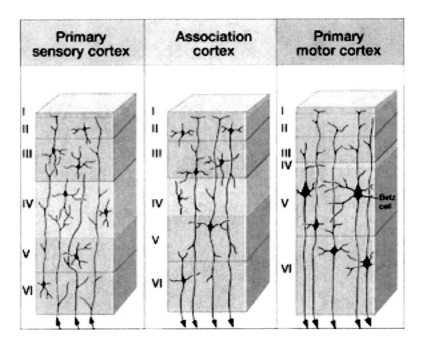

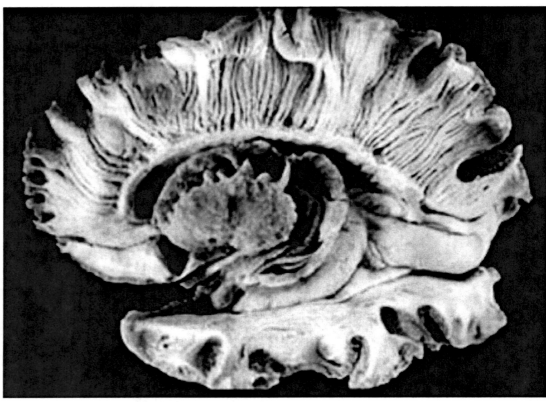

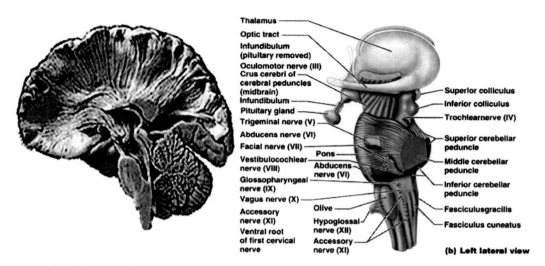

Thalamus
Optic tract
Infundibulum (pituitary removed)
Oculomotor nerve (III)
Crus cerebri of cerebral peduncles (midbrain)
Infundibulum
Pituitary gland
Trigeminal nerve (V)
Abducens nerve (VI)
Facial nerve (VII)
Vestibulocochlear nerve (VIII)
Glossopharyngeal nerve (IX)
Vagus nerve (X)
Accessory nerve (XI)
Ventral root of first cervical nerve

Pons
Abducens nerve (VI)
Olive
Hypoglossal nerve (XII)
Accessory nerve (XI)

Superior colliculus
Inferior colliculus
Trochlearnerve (IV)
Superior cerebellar peduncle
Middle cerebellar peduncle
Inferior cerebellar peduncle
Fasciculusgracilis
Fasciculus cuneatus

(b) Left lateral view

(Left/below) The Corona Radiata. The human Brainstem & Thalamus (Right)

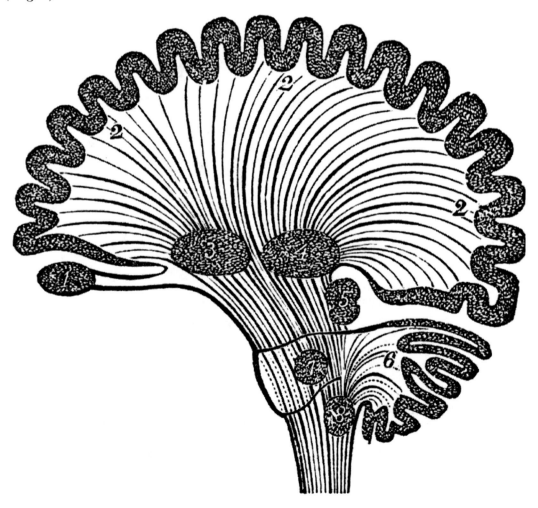

Primary Receiving Areas

Visual input is transmitted from the eyes to the midbrain and thalamus and is transferred to the primary visual receiving area maintained in the neocortex of the occipital lobe (Casagrande & Joseph 1978, 1980; Joseph and Casagrande, 1978).The occipital lobe visual cortex performs complex visual analysis (Montemurro et al 2008; Priebe & Ferster 2008; Schwarzlose et al 2008).

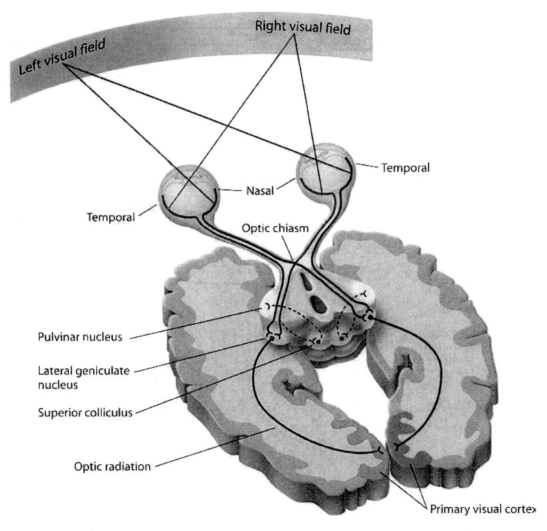

Auditory input is transmitted from the inner ears to the brainstem, midbrain, and thalamus, and is transferred to the primary auditory receiving area within the neocortex of the temporal lobe. The anterior and superior temporal lobes are concerned with complex auditory and linguistic functioning and the comprehension of language (Binder et al., 2010; Lindenberg, R., & Scheef 2007; Yu et al., 2011).

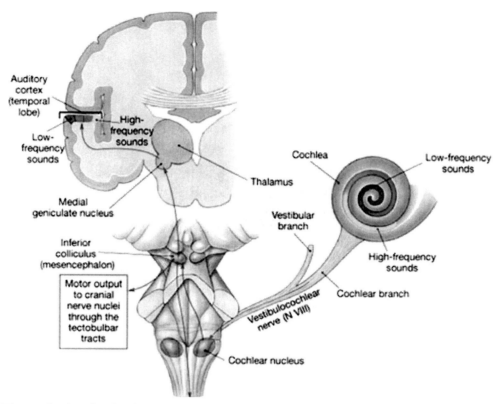

Tactual-physical stimuli are also transmitted from the thalamus to the primary somatosensory areas maintained in the neocortex of the parietal lobe. The parietal lobes are the senior executive of the physical-body-in-space and maintains one's personal image of the body, both physical and visual, and selectively attends to meaningful objects within grasping distance (Billington et al., 2010; Bisley & Goldberg 2010; Chong et al. 2010; Daprati et al., 2010; Gottlieb & Snynder 2010; Nelson et al. 2010).

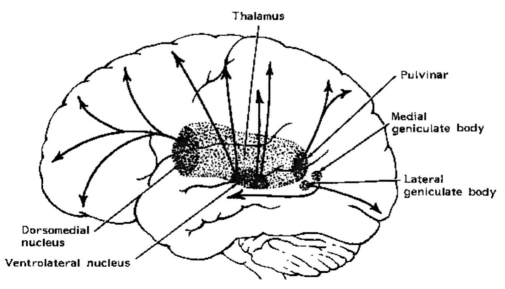

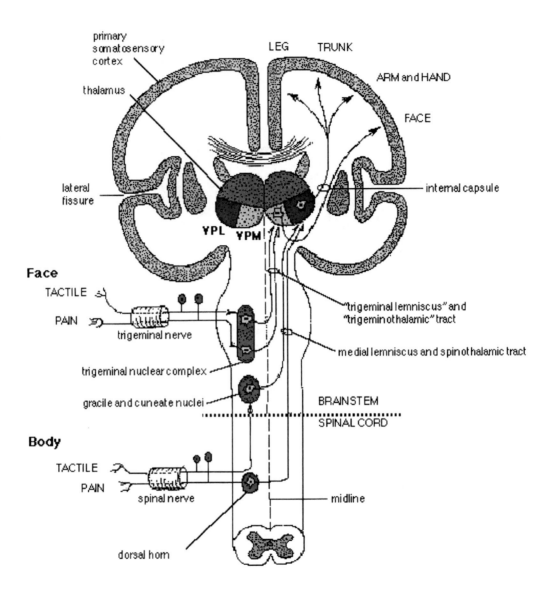

From the primary areas these signals are transferred to the adjoining "association" areas, and simple percepts become more complex by association (Joseph, 1996). Thus, sounds become meaningful and may be perceived as language, and complex visual and tactile stimuli take shape and form and can be recognized as faces, cars, dogs and cats, or a comb or brush.

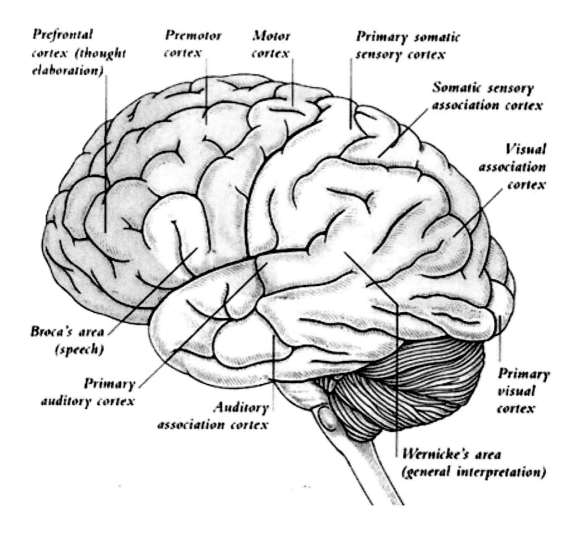

Prefrontal cortex (thought elaboration)   *Premotor cortex*   *Motor cortex*   *Primary somatic sensory cortex*

Somatic sensory association cortex

Visual association cortex

Broca's area (speech)

Primary auditory cortex

Auditory association cortex

Primary visual cortex

Wernicke's area (general interpretation)

Frontal Lobe Monitoring of Activity

Monitoring all this perceptual and sensory activity within the thalamus and neocortex is the frontal lobes of the brain, also known as the senior executive of the brain and personality (Joseph 1986b, 1999a; Joseph et al., 1981). It is the frontal lobes which maintain the focus of attention and which can selectively inhibit any additional processing of signals received in the primary areas.

There are two frontal lobes, a right and left frontal lobe which communicate via a bridge of nerve fibers. Each frontal lobe, and subdivisions within each are concerned with different types of mental activity (Fuster 2008; Joseph, 1999a; Stuss 2009).

The left frontal lobe, among its many functions, makes possible the ability to speak. It is associated with the verbally expressive, speaking aspects of consciousness. However, there are different aspects of consciousness associated not only with the frontal lobe, but with each lobe of the brain and its subdivisions (Joseph, 1986b; 1996, 1999a).

Knowing Yet Not Knowing: Disconnected Consciousness

Consider the well known phenomenon of "word finding difficulty" also known as "tip of the tongue." You know the word you want (the "thingamajig") but at the same time, you can't gain access to it. That is, one aspect of consciousness knows the missing word, but another aspect of consciousness associated with talking and speech can't gain access to the word. The mind is disconnected from itself. One aspect of mind knows, the other aspect of mind does not.

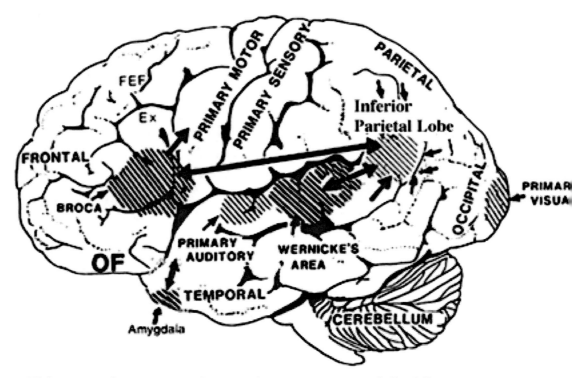

This same phenomenon, but much more severe and disabling, can occur if the nerve fiber pathway linking the language areas of the left hemisphere are damaged. For example, Broca's area in the frontal lobe organizes words received from the posterior language areas, and expresses humans speech. Wernicke's area in the temporal lobe comprehends speech. The inferior parietal lobe sits at the junction of the visual (occipital), somesthetic (parietal) and auditory (temporal) areas and in association with the frontal lobes and Broca's area, associates and assimilates these associations so that, for example, we can say the word "dog" and come up with the names of dozens of different breeds and then visualize and describe them (Joseph, 1982; Joseph and Gallagher 1985; Joseph et al., 1984). Therefore, if Broca's area is disconnected from the posterior language areas, one aspect of consciousness may know what it wants to say, but the speaking aspect of consciousness will be unable to gain access to it and will have nothing to say; called "conduction aphasia."

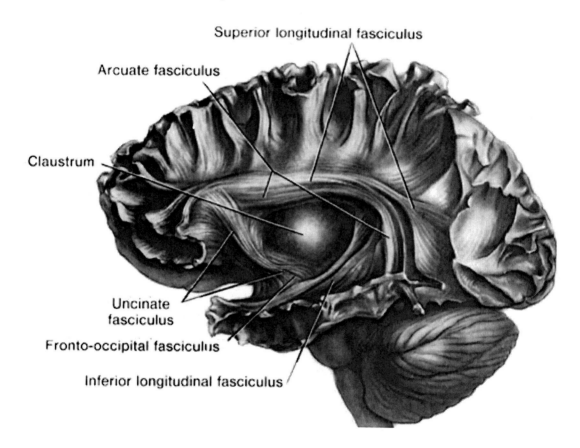

Or consider damage which disconnects the parietal lobe from Broca's area. If you place an object, e.g., a comb, out-of-sight, in the person's right hand, and ask them to name the object, the speaking aspect of consciousness may know something is in the hand, but will be unable to name it. However, although they can't name it, and can't guess if shown pictures, if the patient is asked to point to the correct object, they will correctly pick out the comb (Joseph, 1996). Therefore, part of the brain and mind may act purposefully (e.g. picking out the comb), whereas another aspect of the brain and mind is denied access to the information that the disconnected part of the mind is acting on.

Thus, the part of the brain and mind which is perceiving and knowing, is not the same as the part of the brain and mind which is speaking. This phenomenon occurs even in undamaged brains, when the multiplicity of minds which make up one of the dominant streams of consciousness, become disconnected and/or are unable to communicate.

The Visual Mind: Denial of Blindness

All visual sensations first travel from the eyes to the thalamus and midbrain. At this level, these visual impressions are outside of consciousness, though we may be aware of them. These visual sensations are then transferred to the primary visual receiving areas and to the adjacent association areas in the neocortex of the occipital lobe. Once these visual impressions reach the neocortex, consciousness of the visual word is achieved. Visual consciousness is made possible by the occipital lobe.

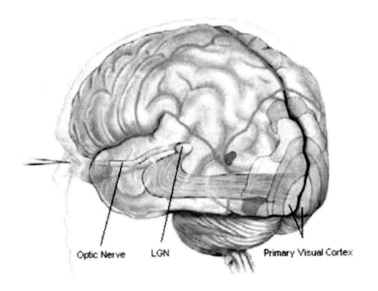

Optic Nerve   LGN   Primary Visual Cortex

Destruction of the occipital lobe and its neocortical visual areas results in cortical blindness (Joseph, 1996). The consciousness mind is blinded and can not see or sense anything except vague sensations of lightness and darkness. However, because visual consciousness is normally maintained within the occipital lobe, with destruction of this tissue, the other mental systems will not know that they can't see. The remaining mental system do not know they are blind.

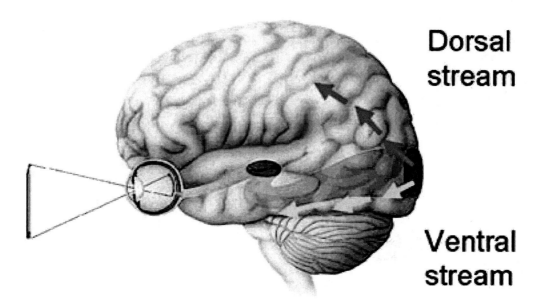

Dorsal stream

Ventral stream

Wernicke's area in the left temporal lobe in association with the inferior parietal lobe comprehends and can generate complex language. Normally, visual input is transferred from the occipital to the inferior parietal lobe (IPL) which is adjacent to Wernicke's area and the visual areas of the occipital lobe. Once these signals arrive in the IPL a person can name what they see; the visual input is matched with auditory-verbal signals and the conscious mind can label and talk about what is viewed (Joseph, 1982, 1986b; Joseph et al., 1984). Talking and verbally describing what is seen is made possible when this stream of information is transferred to Broca's area in the left frontal lobe (Joseph, 1982, 1999a). It is Broca's area which speaks and talks.

Therefore, with complete destruction of the occipital lobe, visual consciousness is abolished whereas the other mental system remain intact but are unable to receive information about the visual world. In consequence, the verbal aspects of consciousness and the verbal-language mind does not know it can't see because the brain area responsible for informing these mental system about seeing, no longer exists. . In fact the language-dependent conscious mind will deny that it is blind; and this is called: Denial of blindness.

Normally, if it gets dark, or you close your eyes, the visual mind becomes conscious of this change in light perception and will alert the other mental realms. These other mental realms do not process visual signals and therefore they must be informed about what the visual mind is seeing. If the occipital lobe is destroyed, visual consciousness is destroyed, and the rest of the brain cannot be told that visual consciousness can't see. Therefore, the rest of the brain does not know it is blind, and when asked, will deny blindness and will make up reasons for why they bump into furniture or can't recognize objects held before their eyes (Joseph, 1986b, 1988a).

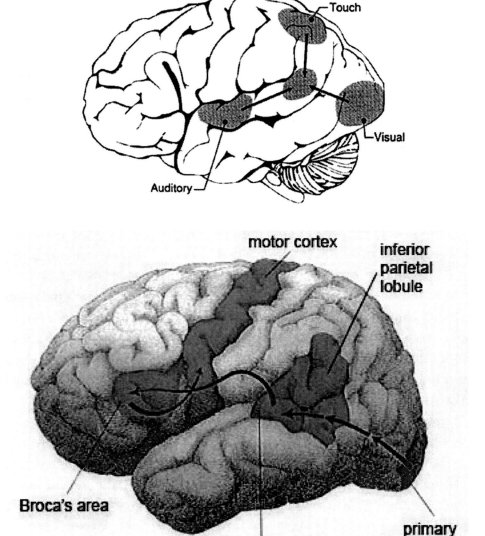

For example, when unable to name objects, they might confabulate an explanation: "I see better at home." Or, "I tripped because someone moved the furniture."

Even if you tell them they are blind, they will deny blindness; that is, the verbal aspects of consciousness will claim it can see, when it can't. The Language-dependent aspects of consciousness does not know that it is blind because information concerning blindness is not being received from the mental realms which support visual consciousness.

The same phenomenon occurs with small strokes destroying just part of the occipital lobe. Although a patient may lose a quarter or even half of their visual field, they may be unaware of it. This is because that aspect of visual consciousness no longer exists and can't inform the other mental realms of its condition.

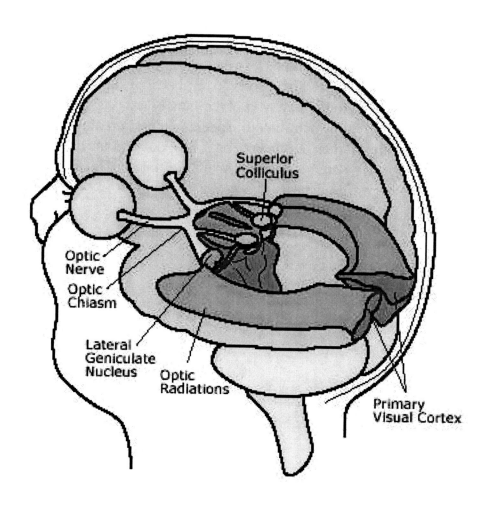

"Blind Sight"

The neocortex is the most recently evolved tissues of the mind. Prior to the evolution of the visual neocortex in the occipital lobe, visual sensations were analyzed primarily by the thalamus and the midbrain superior colliculus. However, even following the evolution of neocortex, visual perceptions continue to be received and analyzed by these more ancient regions of the brain (Casagrande & Joseph 1978, 1980; Joseph & Casagrande 1978, 1980; Van Essen 2005). In consequence, even with total occipital lobe neocortical destruction, the patient may demonstrate a non-conscious awareness of the visual world, a condition referred to as "blind sight" (Cowey, 2010; Persaud et al 2011). Blindsight, in part, is due to the fact that the more ancient subcortical regions of the brain can still see (Joseph 2009), and there remains a non-conscious awareness of visual perceptions which may be transmitted to neocortical regions such as the frontal lobes, and areas 18 and 19 (Anders et al. 2009; Persaud et al 2011).

The brains of reptiles, amphibians, and fish do not have neocortex. Visual input is processed in the midbrain and thalamus and other old-brain areas as these creatures do not possess neocortex or lobes of the brain. In humans, this information is also received in the brainstem and thalamus and is then transferred to the newly evolved neocortex.

As is evident in non-mammalian species, these creatures can see, and they are aware of their environment. They possess an older-cortical (brainstem-thalamus) visual awareness which in humans is dominated by neocortical visual consciousness.

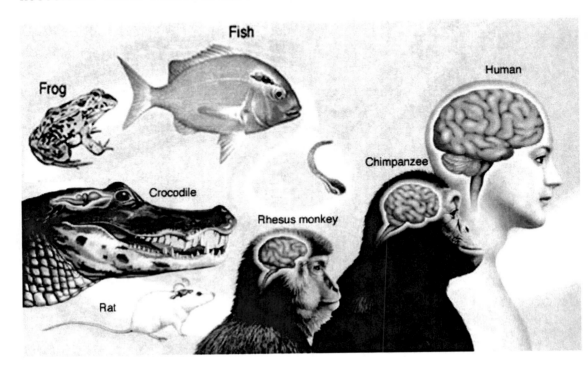

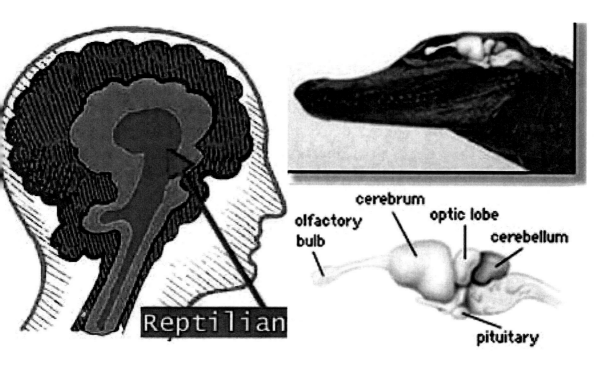

Therefore, even with complete destruction of the visual neocortex, and after the patient has had time to recover, some patients will demonstrate a non-conscious awareness of their visual environment and demonstrate blindsight. Although they are cortically blind and can't name objects and stumble over furniture and bump into walls, they may correctly indicate if an object is moving in front of their face, and they may turn their head or even reach out their arms to touch it--just as a frog can see a fly buzzing by and lap it up with its tongue. Although the patient can't name or see what has moved in front of his face, he may report that he has a "feeling" that something has moved.

Frogs do not have neocortex and they do not have language, and can't describe what they see. However, humans and frogs have old cortex that process visual impressions and which can control and coordinate body movements. Therefore, although the neocortical realms of human consciousness are blind, the mental realms of the old brain can continue to see and can act on what it sees; and this is called: Blind sight (Joseph, 1996).

Body Consciousness: Denial of the Body, and Phantom Limbs

All tactile and physical-sensory impressions are relayed from the body to the brainstem and the thalamus, and are then transferred to the primary receiving and then the association area for somatosensory information

37

located in the neocortex of the parietal lobe (Joseph, 1986b, 1996). The parietal lobes are the senior executive of the physical-body-in-space and maintains one's personal image of the body, both physical and visual, and selectively attends to meaningful objects within grasping distance (Billington et al., 2010; Bisley & Goldberg 2010; Chong et al. 2010; Daprati et al., 2010; Gottlieb & Snynder 2010; Nelson et al. 2010). The parietal lobes receives distinct sensory impressions from the entire body and can feel "pain" or a bug crawling on one's arm, leg, or face (Cohen & McCabe 2010; Daprati et al., 2010; Tsakiris 2010). The entire image of the body is represented in the parietal lobes (the right and left half of the body in the left and right parietal lobe respectively), albeit in correspondence with the sensory importance of each body part. Therefore, more neocortical space is devoted to the hands and fingers than to the elbow.

It is because the body image and body consciousness is maintained in the parietal area of the brain, that victims of traumatic amputation and who lose an arm or a leg, continue to feel as if their arm or their leg is still attached to the body. This is called: phantom limbs. They can see the leg is missing, but they feel as if it is still there; body-consciousness remains intact even though part of the body is missing (Joseph, 1986b, 1996). They may also continue to periodically experience the pain of the physical trauma which led to the amputation, and this is called "phantom limb pain."

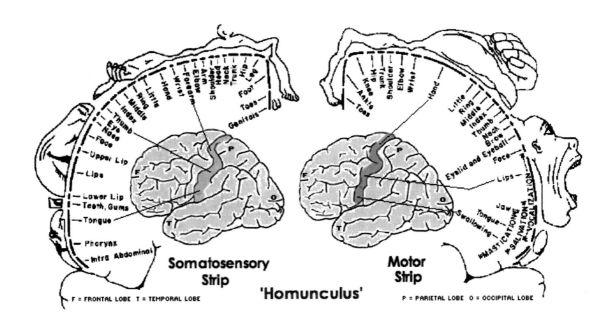

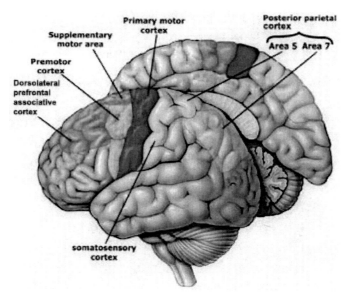

Thus, via the mental system of the parietal lobe, consciousness of what is not there, may appear to consciousness as if it is still there. This is not a hallucination. The image of the body is preserved in the brain and so to is consciousness of the body; and this is yet another example of experienced reality being a manifestation of the brain and mind. In this regard, reality is literally mapped into the brain and is represented within the brain, such that even when aspects of this "reality" are destroyed and no longer exists external to the brain, it nevertheless continues to be perceived and experienced by the brain and the associated realms of body-consciousness.

Conversely, if the parietal lobe is destroyed, particularly the right parietal lobe (which maintains an image of the left half of the body), half of the body image may be erased from consciousness (Joseph, 1986b, 1988a). The remaining realms of mind will lose all consciousness of the left half of the body, which, in their minds, never existed.

Doctor: "Give me your right hand!" (Patient offers right hand). "Now give me your left!" (The patient presents the right hand again. The right hand is held.) "Give me your left!" (The patient looks puzzled and does not move.) "Is there anything wrong with your left hand?"

Patient: "No, doctor."

Doctor: "Why don't you move it, then?"

Patient: "I don't know."

Doctor: "Is this your hand?" (The left hand is held before her eyes.)

Patient: "Not mine, doctor."

Doctor: "Whose hand is it, then?"

Patient: "I suppose it's yours, doctor."

Doctor: "No, it's not; I've already got two hands. look at it carefully." (The left hand is again held before her eyes.)

Patient: "It is not mine, doctor."

Doctor: "Yes it is, look at that ring; whose is it?" (Patient's finger with marriage ring is held before her eyes)

Patient: "That's my ring; you've got my ring, doctor. You're wearing my ring!"

Doctor: "Look at it—it is your hand."

Patient: "Oh, no doctor."

Doctor: "Where is your left hand then?"

Patient: "Somewhere here, I think." (Making groping movements near her left shoulder).

Because the body image has been destroyed, consciousness of that half of the body is also destroyed. The remaining mental systems and the language-dependent conscious mind will completely ignore and fail to recognize their left arm or leg because the mental system responsible for consciousness of the body image no longer exists. If the left arm or leg is shown to them, they will claim it belongs to someone else, such as the nurse or the doctor. They may dress or groom only the right half of their body, eat only off the right half of their plates, and even ignore painful stimuli applied to the left half of their bodies (Joseph, 1986b, 1988a).

However, if you show them their arm and leg (whose ownership they deny), they will admit these extremities exists, but will insist the leg or arm does not belong to them, even though the arm or the leg is wearing the same clothes covering the rest of their body. Instead, the language dependent aspects of consciousness will confabulate and make up explanations and thus create their own reality. One patient said the arm belonged to a little girl, whose arm had slipped into the patient's sleeve. Another declared (speaking of his left arm and leg), "That's an old man. He stays in bed all the time."

One such patient engaged in peculiar erotic behavior with his left arm and leg which he believed belonged to a woman. Some patients may develop a dislike for their left arms, try to throw them away, become agitated when they are referred to, entertain persecutory delusions regarding them, and even complain of strange people sleeping in their beds due to their experience of bumping into their left limbs during the night (Joseph, 1986b, 1988a). One patient complained that the person sharing her bed, tried to push her out of the bed and then insisted that if it happened again she would sue the hospital. Another complained about "a hospital that makes people sleep together." A female patient expressed not only anger but concern least her husband should find out; she was convinced it was a man in her bed.

The right and left parietal lobes maintain a map and image of the left and right half of the body, respectively. Therefore, when the right parietal lobe is destroyed, the language-dependent mental systems of the left half of the brain, having access only to the body image for the right half of the body, is unable to become conscious of the left half of their body, except as body parts that they then deduce must belong to someone else.

However, when the language dominant mental system of the left hemisphere denies ownership of the left extremity these mental system are in fact telling the truth. That is, the left arm and leg belongs to the right not the left hemisphere; the mental system that is capable of becoming conscious of the left half of their body no longer exist.

When the language axis (Joseph, 1982, 2000), i.e. the inferior parietal lobe, Broca's and Wernicke's areas, are functionally isolated from a particular source of information, the language dependent aspect of mind begins to make up a response based on the information available. To be informed about the left leg or left arm, it must be able to communicate with the cortical area (i.e. the parietal lobe) which is responsible for perceiving and analyzing information regarding the extremities. When no message is received and when the language axis is not informed that no messages are being transmitted, the language zones instead relie on some other source even when that source provides erroneous input (Joseph, 1982, 1986b; Joseph et al., 1984); substitute material is assimilated and expressed and corrections cannot be made (due to loss of input from the relevant knowledge source). The patient begins to confabulate. This is because the patient who speaks to you is not the 'patient' who is perceiving- they are in fact, separate; multiple minds exist in the same head.

Summary: Brain & Mind.

The brain and the mind are synonymous and are hierarchically, vertically, and horizontally organized, with different regions engaging in both parallel and localized processing, with specialized neurons performing discrete functions, such as face recognition, and others participating in large neural networks to make complex cognitive functioning processing possible, including, for example, language and consciousness.

There are tissues of the mind which mediate language and complex cognition, such as the neocortex and lobes of the brain, those aspects associated with sex, love and war, such as the limbic system and those areas which respond reflexively and completely unconsciously, laboring to maintain homeostasis, and the rhythmics of waking, sleeping, dreaming, and of the heart, the respiratory systems, and the vegetative mind. These latter unconscious activities are associated with the functional integrity of the brainstem, i.e., the medulla, pons, and midbrain, and to a much lesser extent, the adjacent diencephalon, i.e. the thalamus and hypothalamus.

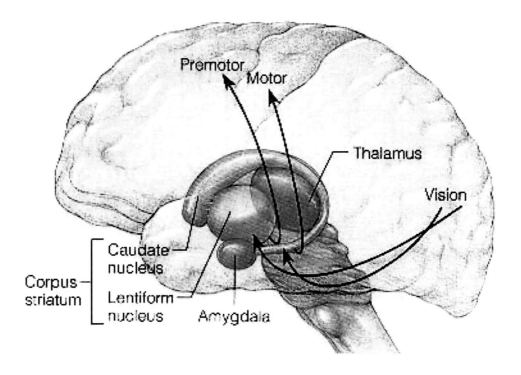

The mind and brain are also functionally lateralized, with specific interactive subareas in the left hemisphere mediating the linguistic, temporal, sequential, arithmetical, and rhythmical aspects of consciousness and verbal thought. It is the right half of the brain which subserves and provides the foundations for an interactive non-verbal awareness encompassing music, emotion, the body, prosodic language, and spatial thought.

Thus, the cerebrum and psyche are organized such that several semi-independent mental system coexist, literally one on top of the other, side by side, distributed and localized, with the limbic system and brainstem providing much of the "food for thought" that is received by the neocortex and reflected upon by the conscious mind.

The Brainstem

The brainstem is exceedingly ancient and complex structure (Aitkin, 1986; Vertes, 1990), and is concerned with sensory reception, arousal, and reflexive movements of the body, including stereotyped and routine motor acts such as yawning, opening and closing the mouth, wiping the face, or lifting the legs, taking steps, walking or running (Aminoff, 1996; Blessing, 1997; Bidelman et al. 2011; Klemm, 1990; Roh et al., 2011; Shen et al., 2011). Although capable of learning, forming simple memories, and displaying synaptic plasticity (Jones & Pons, 1998; Isukahara, 1985; Moore & Aitkin, 1975; Moore & Irvine, 1981), this portion of the brain cannot think, reason, or feel love or sorrow (Joseph, 1999c).

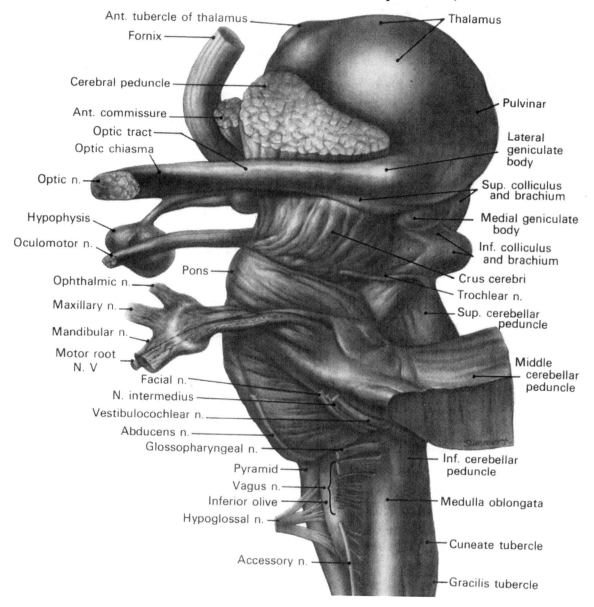

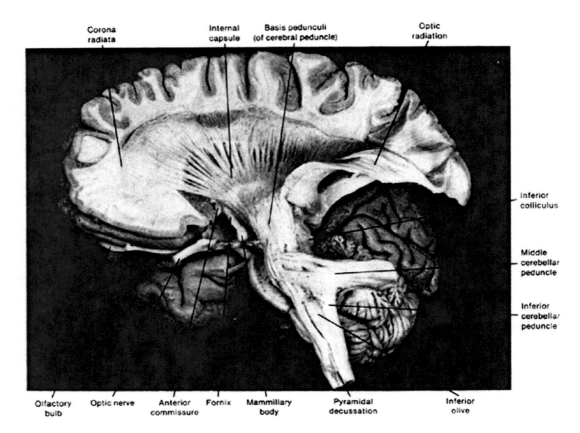

The brainstem is both diffusely organized and functionally specialized. When specific brainstem nuclei (and associated neural networks) are stimulated, sucking, chewing, swallowing, swimming, stepping, walking, and running movements can be induced (Bidelman et al. 2011; Klemm, 1990; Mogenson, 1990; Steriade & McCarley, 2005; Skinner & Garcia-Rill, 1990; Vertes, 1990). HoweRoh et al., 2011; Shen et al., 2011). Moreover, these movements are hierarchically organized with those requiring the least amount of organizational or conscious control being localized near the medulla-spinal border. It is noteworthy, however, that this areas also contains neurons which when stimulated can trigger female sexual posturing (Benson, 1988; Rose, 1990); i.e. the lordosis (or "doggie") position. These latter neurons are interconnected with the amygdala and ventromedial hypothalamus--nuclei which are also involved in sexual activities. At the level of the spinal cord, movement programs are exceedingly simplified and usually consist of only fragments of the entire motor display.

As one ascends from the spinal cord to the medulla, then the pons, and finally the midbrain, the degree, extent, and nature of these various motor programs and behaviors becomes increasingly complex. For example, when the most caudal regions of the medulla are stimulated stepping motions can be induced, whereas stimulation of the midbrain can initiate eye movements, head turning, controlled walking, running (Cowie & Robinson 1994; Skinner & Garcia-Rill, 1990) and vocalization--which is produced by the periaqueductal gray (Jurgens, 2009; Zhang et al. 1994).

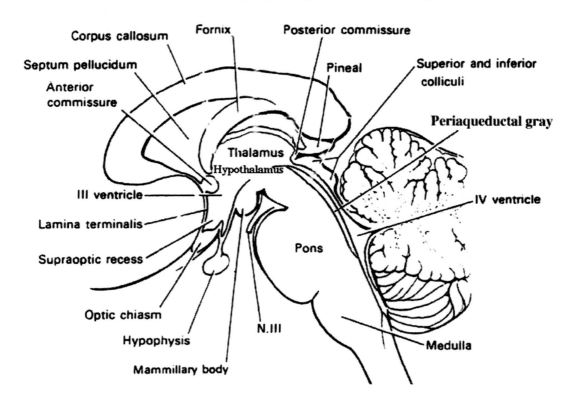

The periaqueductal gray coordinates the activity of the laryngeal, oral-facial, and principal and accessory muscles of respiration and inspiration (Zhang et al. 1994) and receives extensive input from the amygdala, the anterior cingulate, and the left and right frontal lobes--structures which are all implicated in vocalization, and/or speech and language. However, it is the coordinated activity of the periaqueductal gray which enables an individual to laugh, cry, or howl, even if the rest of the brain (excepting the brainstem) were dead.

The brainstem is devoid of cognitive activity as it is designed to react immediately and reflexively to sensory stimuli. Consider, for example, infants born with only a brainstem, i.e. anencephalics, who are completely devoid of any semblance of conscious or cognitive activity; and the same is true of adults who suffer a brainstem-forebrain transection. Although they live and breath, these latter unfortunate souls are essentially forebrain dead.

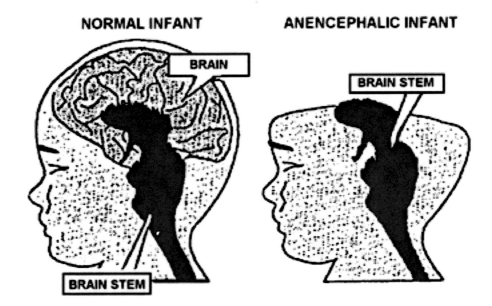

NORMAL INFANT   ANENCEPHALIC INFANT

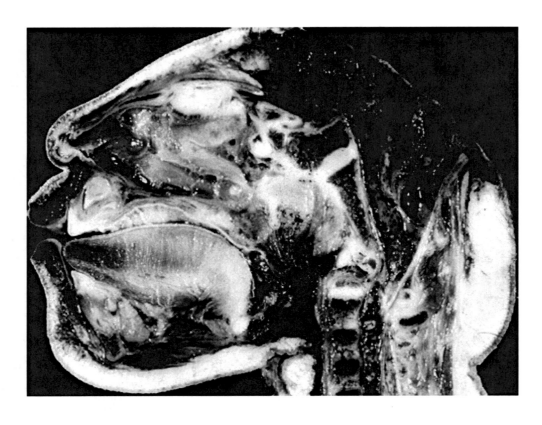

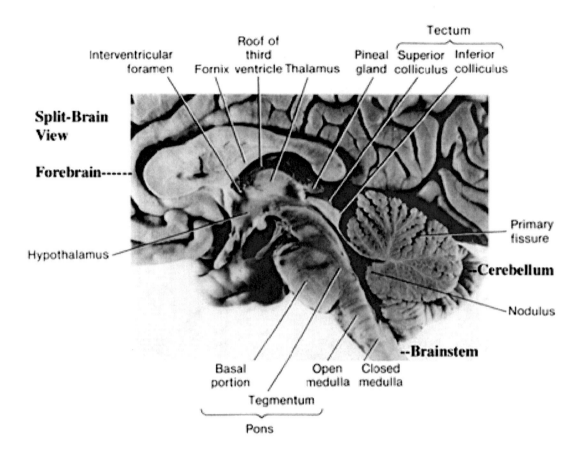

The brainstem, although almost wholly unconscious, maintains consciousness and promotes cognitive activity by feeding the forebrain with activating neurotransmitters such as norepinephrine (NE) Serotonin (5HT), and dopamine (DA) and by controlling arousal through the reticular activating system (Blessing, 1997; Halbach & Demietzel 2006; Steriade & McCarley, 2005; Usher et al., 1999). Because of its importance in these and other life-preserving functions, the stereotypical consequences of brainstem injury, therefore, are prolonged unconscious and death.

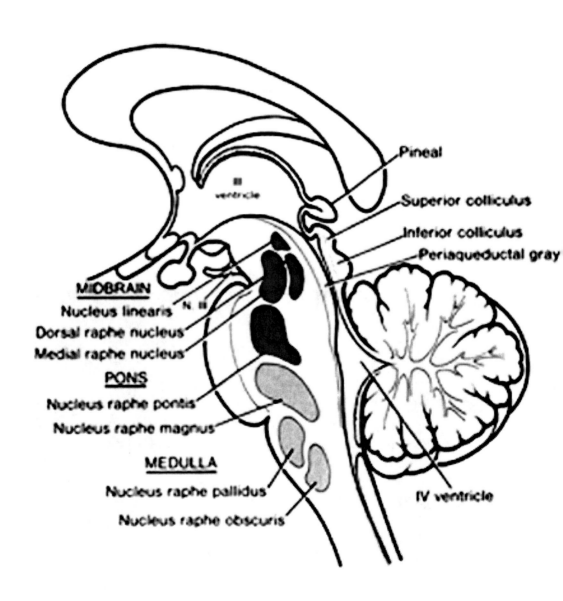

Brainstem Cranial Nerves

With the exception of vision and olfactory sensations, all sensory information is relayed from the spinal cord and cranial nerves to the brainstem. From the brainstem this information is transmitted to the thalamus. Conversely, all motor commands, be it the movement of the lips, the tapping of a finger, or the blink of an eye, must be transmitted to the brainstem cranial nerves and/or spinal cord before they can be carried out. Cranial nerves also influence breathing and heart rate.

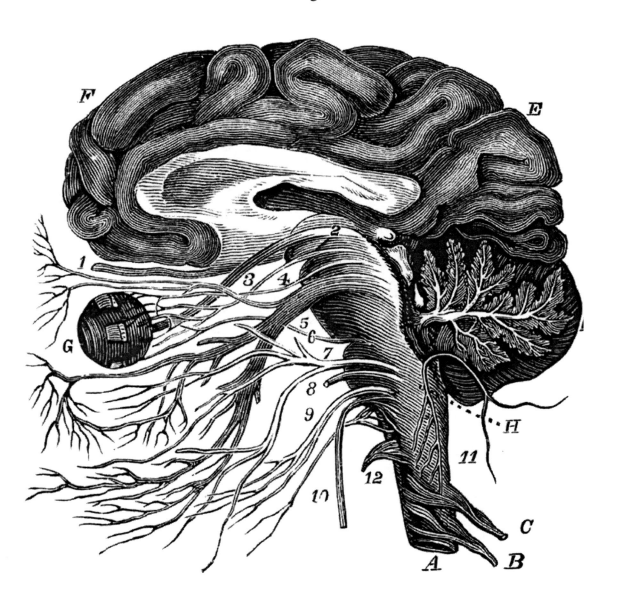

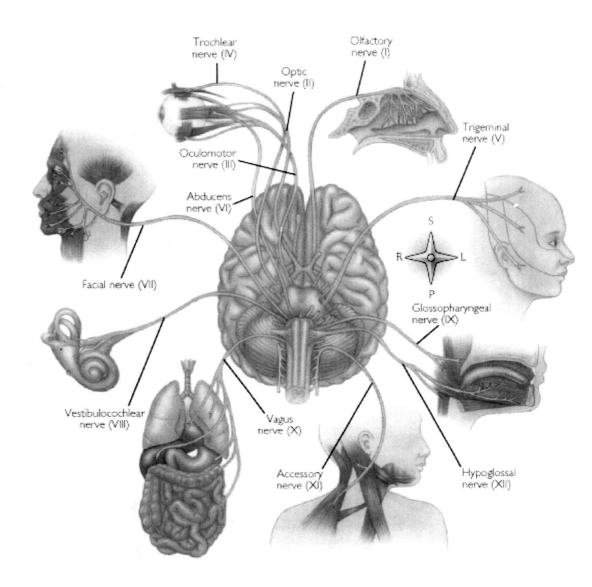

CRANIAL NERVE 12: HYPOGLOSSAL

The hypoglossal nucleus located in the caudal medulla gives rise to the 12th nerve which controls tongue movements via innervation of the somatic skeletal muscles. A lesion, tumor, or infarct invading this area can give rise to weakness or atrophy of the tongue. For example, if the tongue is weak and deviates to the right, the ipsilateral lower motor neuron (LMN) is involved, indicating that the right genionglussus muscle is weak. Weakness can be tested by having the patient place his tongue in one side of cheek and ask him to press against your finger (on the outside of cheek). The clinical and EMG signs of an LMN lesion of the 12th nerve are hemiatrophy, unilateral fasiculation or fibrillation, and usually severe paralysis with obvious deviation to the paralytic side when the tongue is protruded.

CRANIAL NERVE 11: SPINAL ACCESSORY

Cranial nerve 11 consists of two distinct segments; a cranial portion which, along with the vagus nerve, forms the inferior laryngeal nerve which innervates the muscles of the larynx; and a spinal portion which innervates the sternocleidomastoid and upper trapezious muscles and therefore aids in turning the head and elevating the shoulders. Injuries to this nerve may result in a sagging of the shoulder on the affected side, or weakness in turning the head -particularly when resistance is applied.

CRANIAL NERVE 10: VAGUS

The vagus nerve is actually a complex mix of nerves which innervate a variety of structures including the trachea, larynx, pharynx, esophagus, and abdominal and thoracic viscera, the epiglottis, and the external auditory meatus. Hence, this nerve (in conjunction with the 9th nerve -both of which are derivatives of the skeletal muscles that originally formed the brachial (gill) arches) is important in regard to oral activities, including swallowing, breathing, and speaking, pharyngeal constriction and thus with movement of the palate, pharynx and larynx. The vagus nerve is therefore responsible for swinging the soft palate upward and backward to contact the posterior wall of the pharynx, sealing off the oropharynx from the nasopharynx when one swallows, whistles, or speaks.

An injury to this region can result in severe and enduring palatal weakness as well as a condition referred to as pseudobulbar palsy. Unless the soft palate elevates properly liquid will escape into the nose when one drinks, and when one speaks resulting in nasal speech. Indeed, damage to these nuclei can severely effect speech.

CRANIAL NERVE 9: GLOSSOPHARYNGEAL

The 9th nerve is closely related to the vagus and both have similar functional components, including general somatic and visceral afferents and efferents. The glossopharyngeal nerve thus receives tactile, thermal, and pain sensations from the tongue, and contributes fibers to the solitary nucleus thereby forming the gustatory nucleus. In addition, the glossopharyngeal nerve receives fibers from the carotid sinus which transmit impulses regarding increases in carotid arterial pressures. The ninth nerve then transmits this data to the solitary nucleus which in turn contributes fibers to the vagus nerve. In this manner the 10th and 9th nerve can induce reductions in arterial blood pressure and heart rate.

Lesions to the 9th nerve generally results in a loss of taste and sensation in the posterior third of the tongue, and a loss of the gag reflex and the carotid sinus reflex. In some cases intense pain may be triggered by swallowing or coughing.

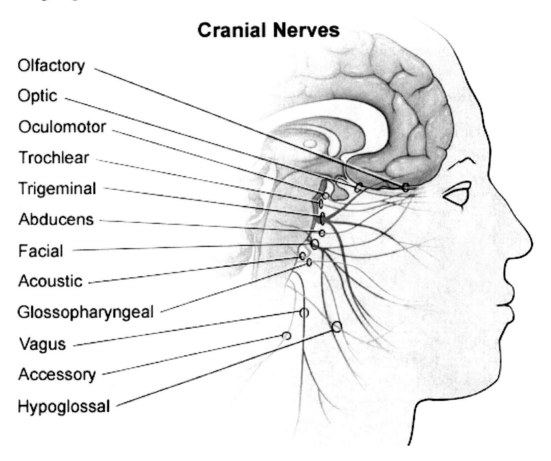

Cranial Nerves

Olfactory
Optic
Oculomotor
Trochlear
Trigeminal
Abducens
Facial
Acoustic
Glossopharyngeal
Vagus
Accessory
Hypoglossal

THE EIGHT CRANIAL NERVE & THE COCHLEAR SYSTEM
TINNITUS, DEAFNESS, DIZZINESS, VERTIGO

A portion of the 8th nerve arises from the cochlear nuclei so as to form the auditory branch of the 8th nerve which in turn bifurcates and innervates the ampulla, utricle, saccule and cochlear duct of the semi circular canals of the inner ear. It is via the acoustic nerve that auditory stimuli are relayed to the ventral and dorsal cochlear nuclei for further transduction prior to transfer to the inferior colliculus of the midbrain, and the medial geniculate nucleus of the thalamus. Hence, lesions involving this nerve or the cochlear nucleus can give rise to significant hearing problems, including deafness and tinnitus --ringing in the ears which may be reported as buzzing, humming, whistling, roaring, hissing, or clicking.

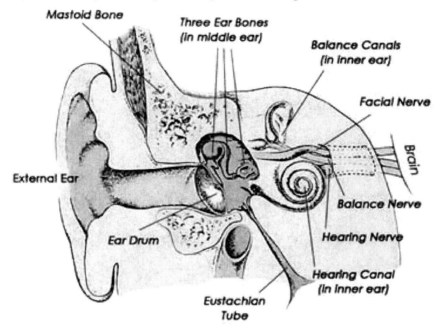

Tinnitus. There are two general types of tinnitus, vibratory, and non-vibratory. Non-vibratory tinnitus is actually quite common, being due to contraction of the muscles of the inner and middle ear, but is usually masked and may only be heard at night or when in a very quiet environment. In contrast, vibratory tinnitus is most often due to disorders of the inner ear, ossicles of the middle ear, tympanic membrane, or 8th nerve, and may be accompanied by deafness in one ear.

Deafness. If a patient complains of deafness, it must be determined if this is due to otosclerosis, chronic otitis, occlusion of the external auditory canal or eustachian tube, in which case the problem is conductive and is not related to nerve damage. By contrast, disease of the cochlear or of the cochlear division of the 8th nerve, or its central connections results in sensorineural (or nerve) deafness.

These two distinct disorder can be easily differentiated with the help of a tuning fork. When struck, it is placed at the center of the forehead or on the mastoid bone behind the ear so that the vibration can bypass the middle ear and can be mechanically conveyed to the inner ear so as to excite auditory impulses.. If the condition is due to nerve deafness, the sound is localized in the normal ear and fails to be perceived by the effected ear. If the disorder is conductive, the sound is heard as louder in defective ear. If there is an obstruction or a middle ear abnormality (vs a nerve disorder) then the sound is heard louder in that ear due to dampening of noise via the air conduction block. If there is a lesion of the nerve then the sound is heard in the opposite of normal ear only.

THE 8th (VESTIBULAR) NERVE

The vestibular division of the 8th nerve arises from the vestibular nerve which innervates the labyrinth and the maculas of the saccule and utricle and the ampullae of the semicircular canals. The vestibular system was derived from the lateral line system, and in turn gave rise to the evolution of the cerebellum.

The major concern of the vestibular division is not hearing or vibratory perception, however, but determination of the body's position in visual-space so as to maintain equilibrium during movement. All sensory receptors for the vestibular system are located in the membranes of the labyrinth of the internal ear, and in the vestibular ganglion which is in the internal auditory canal. Thus it can determine changes in position via alterations in fluid balance within the semi-circular canals.

Lesions of the vestibular receptors, their nerves, or central connections cause abnormal sensations of movement, vertigo, nausea, tendencies to fall, dizziness, and motion sickness, etc.

The vestibular system also provides information about the position of the head and correlates head and eye movements with somatic muscle activity. Together with the descending medial longitudinal fasciculus (MLF) and vistibulospinal tracts, the vestibular system is able to mediate the postural reflexes. In part it is able to accomplish this via rich interconnections with those cranial nerve nuclei (via the MLF) which subserve eye movement (nerves VI, IV, & III). Disease involving these tissues can therefore cause nystagmus.

Following injuries to this system, patients may complain of to-and-fro or up-and-down movements of body, or floors, and noted that the walls seem to tilt or sink or rise. When walking there may be feelings of unsteadiness such that they veer to one side. Or there may be a feeling of being pulled or drawn--a feeling of impulsion. There also may be a disinclination to walk (particularly during an attack), a tendency to list to one side, and the condition may be aggravated by riding in a vehicle. Some disturbances may occur only for a few seconds, or after lying down or sitting up, turning, etc. When less severe the patient may merely veer to one side while walking.

The vestibular system is also concerned with eye movement, for it is also via ocular signals that the position of the head and body in space can be determined. Hence, vestibular dysfunction can include difficulty focusing or fixating on objects while walking, or when the object is moving. This is due to a loss of stabilization of ocular fixation by the vestibular system during body movement and is caused by an inability to integrate visual with vestibular input. These functions are normally made possible through rich interconnections (via the MLF) between the vestibular system and the 6th, 4th and 3rd nerves which subserve eye movement, as well as the bilateral interconnections between these regions and the cerebellum.

ORAL-FACIAL MOVEMENT & SENSATION
THE 7th CRANIAL (FACIAL) NERVE

The 7th nerve, and the brainstem nuclei which it innervates, is concerned with facial movement, including elevation of the eyebrows, retraction of the lips, closure of the auditory canals, as well as with gustatory sensation. Injuries to the 7th nerve can therefore produce a lip retraction, eyebrow lifting or eyelid closure paralysis; i.e. Bell's palsy. Patients have difficulty or are unable to wrinkle the forehead, purse their lips and show their teeth, and the corner of the mouth may droop.

In addition, whereas cranial nerves 9 and 10 innervate the taste buds of the posterior 3rd of the tongue, cranial nerve 7 innervates the anterior 2/3s. Hence, a lesion of the 7th nerve can result a disturbance of taste sensation.

In addition to the muscles of the face, the 7th nerve also innervates the stapedius muscle, which acts to dampen excessive sound via inhibition of the movement of the ossicles. If the 7th nerve is injured, the stepedious may become paralyzed and the patient may report that sounds are uncomfortably loud. However, disturbances of hearing are most usually associated (at least at the brainstem level) with damage involving the 8th cranial nerve.

As to oral and facial movement including swallowing and speaking, the functional integrity and participation of the 5th nerve is also important for it controls the jaw. In this regard, the 5th, 7th, 9th, 10th, and 12th nerve and associated nuclei frequently act in concert regarding oral-facial, jaw, head and shoulder movement, and are richly interconnected.

THE 5th CRANIAL NERVE: TRIGEMINAL

The 5th nerve is the largest of the cranial nerves and innervates the trigeminal nucleus within the medulla. It is concerned with jaw closure, as well as chewing, grinding, and lateral movement of the jaw. In this regard the 5th nerve also acts in concert with the 7th nerve which innervates all the muscles involved in facial expression. These muscles control the size of every facial aperture, including the auditory canals. A lesion involving this nucleus, therefore, can result in difficulty chewing, and if severe, atrophy and complete paralysis of the left or right temporal and masseter muscles. Paralysis and atrophy are always ipsilateral to the lesion.

The general somatic afferent components of the 5th nerve also mediate the general sensory modalities for the face, teeth, and mouth, and the mucous membranes of the nose, check, tongue, and sinuses. These general sensory functions include proprioception, touch, pain, and temperature.

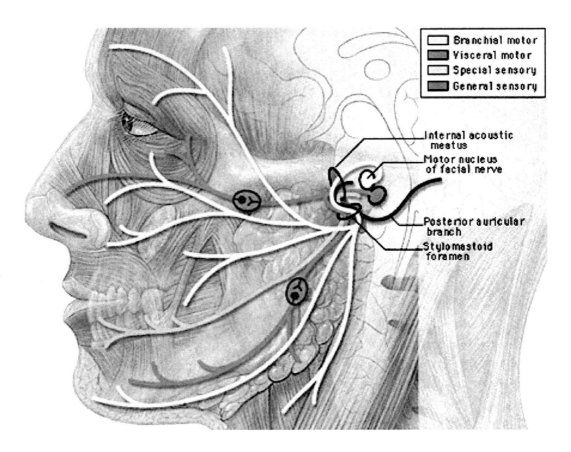

EYE MOVEMENT: THE 6th CRANIAL NERVE: ABDUCENS

The abducens is a motor nerve that innervates the lateral rectus muscle of the eye, and is part of a collection of fibers which are found with the loop of fibers that also form the facial nerve. However, fibers from the abducens nuclei also ascend the brainstem and terminate on neurons belonging to the oculomotor complex which innervates the medial rectus muscle. It is therefore responsible for horizontal eye movements. In large part, the pontine center for lateral gaze and the abducens nuclei form a single entity (Carpenter 1991). In addition, the abducens is linked to the pontine/midbrain center for vertical gaze (the rostral interstitial nucleus of the MLF).

Thus abducens nerve and nuclei are the pontine centers which control lateral eye movements outward to the right or to the left. Injuries to the 6th nerve can therefore produce a lateral gaze paralysis as well as a paralysis of the lateral rectus muscle which results in double vision (horizontal diplopia).

CRANIAL NERVES OF THE MIDBRAIN EYE MOVEMENT

Movement of the eye is dependent on the functional integrity of the posterior parietal neocortex, the frontal eye fields of the lateral frontal lobes, the midbrain visual colliculi, cerebellum, and the cranial nerves and nuclei of the upper pons (cranial nerve VI) and midbrain (nerves IV & III). All pathways mediating saccadic, pursuit, and vestibulocular movements, including those originating in the forebrain and midbrain, then converge onto the pontine centers for horizontal gaze.

CRANIAL NERVE 4: TROCHLEAR

The 6th (abducens) nerve and nuclei are the pontine centers which control lateral eye movements outward to the right or to the left. By contrast, the 4th (trochlear) nerve and nucleus is located just caudal to the inferior colliculi, and innervates the superior oblique muscle of the eye; a muscle which has three actions: depression, abduction, intorsion. The 4th nerve therefore assists in moving the eye downward or inward.

CRANIAL NERVE 3: OCULOMOTOR

The 3rd (oculomotor) nerve is responsible for rotating the eye upward, downward, as well as inward and innervates all ocular rotary muscles (except for the lateral rectus and superior oblique). These include the medial, superior, inferior recti and inferior oblique.

The 3rd never also innervates the intraocular and smooth muscles of the pupil; i..e. the ciliary and pupilloconstrictor muscles. Lesions, therefore, may result in an inability to rotate the eye upward, downward, or inward, and the pupil (on the side of the lesion) may fail to respond to direct light.

In addition, the 3rd nerve innervates the levator palprebrae muscle which elevates the eyelid. When the third nerve has been injured this may result in levator palpebrae weakness and thus ptosis of the eyelid.

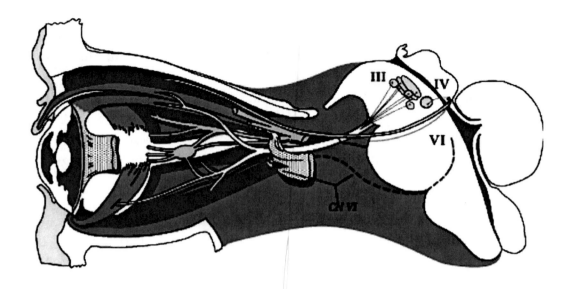

THE OPTIC NERVE

The optic nerves are responsible for transmitting all visual impulses form the retina to the brain, first crossing over at the optic chasm, and then penetrating the brain to form the optic tracts which terminate in the lateral geniculate nucleus of the thalamus. Specifically, beginning at the retina, the rods and cones project to horizontal and bipolar cells which in turn project to X, Y and W ganglion cells within the retina. Axons of the retinal ganglion cells run in parallel along the surface of the retina and converge at the optic disc (the blind spot) where they gather together and punch through the retina thereby forming the optic nerve.

The right and left optic nerves project into the cranial cavity via the optic foramina and then unite to form the optic chiasm where a decussation occurs, thereby forming the optic tracts. Subsequently, visual input from the left half of visual space projects to the right hemisphere, and input from the right half of visual space projects to the left cerebrum via the optic tracts which in turn project predominantly to the lateral geniculate nucleus of the thalamus, and to a much lesser extent, to the superior colliculus, and probably the hypothalamus, amygdala, and inferior temporal and superior parietal lobes. Because the optic nerve does not project directly to the brainstem, it is not a true cranial nerve -at least in mammals- although some fibers (via the optic tract) are received in the midbrain visual colliculi.

From the lateral geniculate nucleus these visual fibers form the optic radiations which project predominantly to the striate cortex within the occipital lobe, and give off collaterals to the inferior temporal and superior parietal lobe where the upper and lower visual fields are also represented.

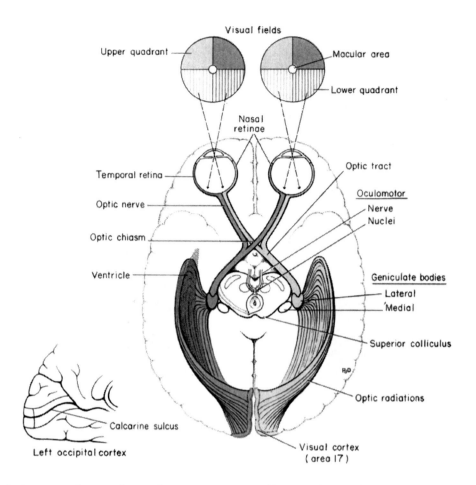

Injury to the optic pathways necessarily produce visual defects and are referred to as homonymous if they are restricted to either the right or left visual field, or heteronymous if both visual fields are disrupted to some degree. Heteronymous defects suggest a lesion involving both cerebral hemispheres, or a lesion to retina or the optic nerve before it crosses over at the decussation.

If the visual defects are homonymous, then the lesion can be localized to one side of the optic tract or radiations and thus to the right or left hemisphere. Complete destruction of the optic tract or its terminal zones induces a homonymous hemianopsia to the left or right, whereas a partial injury may induce a quadratic homonymous defect: upper quadrant being associated with temporal lobe defects, and lower quadrant defects associated with the superior parietal lobe.

THE OLFACTORY NERVE

The olfactory nerve also bypasses the brainstem, and is not a true cranial nerve, but a complex axonal pathway which projects directly to a wide variety of forebrain structures, including the amygdala, entorhinal cortex, hypothalamus, orbital frontal lobes, and the dorsal medial nucleus. Nevertheless, the olfactory nerve is associated with brainstem functions as originally it was via the olfactory system that information regarding not just smell but taste were derived. That is, early in the course of evolution, the olfactory pathway split apart and became segregated (perhaps with the emergence of the vomeronasal organ) so as to form distinct pathways that subserve smell and taste. Nevertheless, smell remains important to taste, which is why when suffering from a severe cold, there is a generalized dampening of the sense of taste.

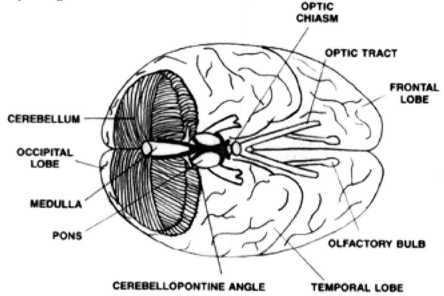

The olfactory nerve originates via bipolar cells located in the olfactory epithelium which in turn is composed of very primitive cells that have a life span that last only days. New axons and new synapses are continually forming. These nerve fibers perforate and pass through the cribriform plate (making them very susceptible to injury and shearing) and project directly to the olfactory bulb. From the olfactory bulb the olfactory tract is formed which projects to pyriform, periamygdaloid, and entorhinal areas -which in turn constitute the primary olfactory cortex. Projections continue to the amygdala, hippocampus, thalamus, orbital frontal lobes, and insula.

Following a head injury the cribriform plate may fracture, the olfactory nerve may be severed, and the meninges may rupture. If this occurs, in addition to loss of smell, cerebrospinal fluid may continually drip or gush into the nose -which suggests a cerebrospinal fluid fistula. To determine if nasal drip is due to CSF leakage or normal mucous secretion, the fluid can be subject to a glucose test tape (as used for urinalysis). CSF contains glucose, whereas mucus does not. If glucose is indicated, call a neurosurgeon. However, in some cases, a head injury may cause these nerves to be severed although the cribriform plate remains intact. In these instances, loss of smell may result: anosmia. However, this loss may be unilateral or bilateral (involving both nostrils). If unilateral, the patient will not notice the loss, in which case each nostril must be assessed separately.

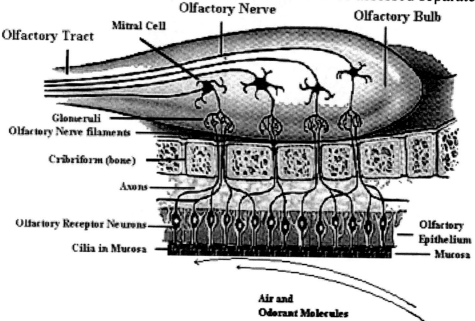

Dysosmia refers to a perversion of the sense of smell, which in turn may be due to partial injuries of the olfactory bulbs, or due to tumor of the nasal sinuses or the temporal lobes in which case food may also be said to have an extremely unpleasant odor and/or taste. Similarly, olfactory hallucinations are associated with tumors, seizure activity, as well as head injuries involving the inferior temporal lobes.

The Cerebellum

The cerebellum sits atop the brainstem and accounts for approximately 25% of the brain. It communicates with almost all regions of the neuroaxis, with the single exception of the striatum, and has been implicated in cognitive, emotional, sensory, motor and speech processing (Almeida & Afonso 2011, Bledsoe et al., 2011; Schmahmann, 1997; Wills et al., 2011) . The cerebellum has been shown to display neuroplasticity (Nimura et al., 1999) and learning and memory (Okamoto et al., 2011; Molinari, et al., 1997) and may well serve as an integrative interface for cognition, emotion, motor functioning and memory ((Ito 2011).

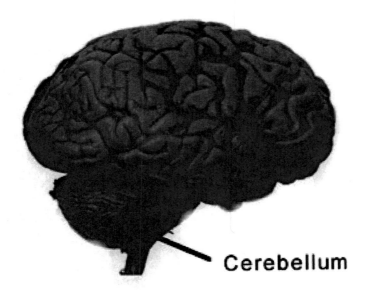

Cerebellum

The cerebellum is typically thought of as a motor center. However, electrical-electrode stimulation or damage to this structure can trigger rage reactions (Bharos, et al. 1981), and hyperactivity (Carpenter, 1959), including "mania" (Cutting, 1976). Abnormalities in the cerebellum have also been implicated in the pathogenesis of schizophrenia and autism (Bauman & Kemper, 1985; Courchesne & Plante, 1996; Gaffeny, et al. 1987; Heath, 1977; Heath, et al. 1979, 1982; Taylor 1991; Weinberger et al. 1979, 1980). Although the notion that abnormal rearing conditions may contribute to autistic and schizophrenic behavior is no longer in fashion, it is noteworthy that Heath (1972) found that monkeys reared under deprived conditions displayed abnormal electrophysiological activity in the cerebellum (dentate gyrus) as well as the septal nuclei. These animals also displayed autistic behavior (Harlow & Harlow, 1965a,b). These findings are significant, for the cerebellum is an outgrowth of the vestibular system, and insufficient social-emotional or physical stimulation would also result in insufficient vestibular activation.

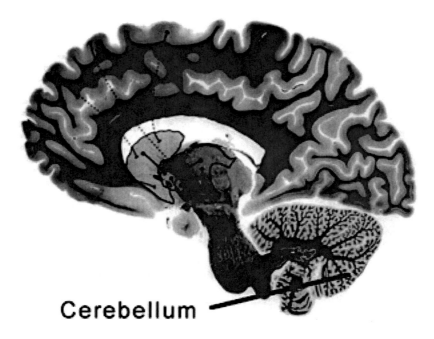

Cerebellum

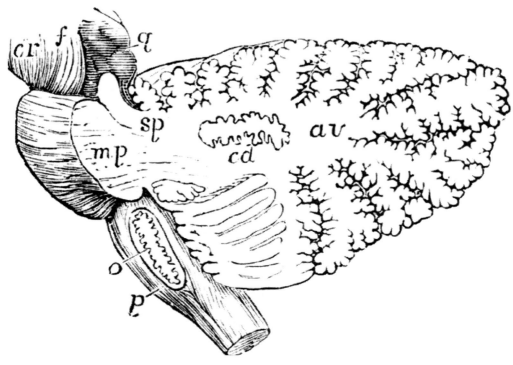

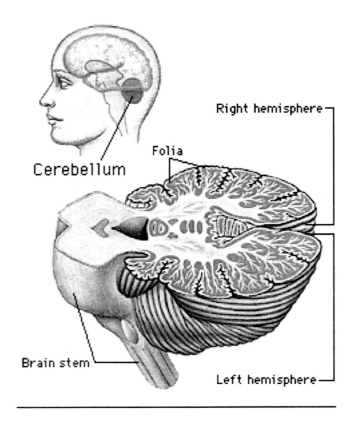

Cerebellum

Folia

Right hemisphere

Brain stem

Left hemisphere

The cerebellum consists of a number of structures and distinct neural circuits associated with specific fiber systems, including climbing, parallel and mossy fibers. It also appears that different regions are concerned with different functions that also have a major motor component. These functions include speech (Silveri et al. 1994; van Dongen et al. 1994; Wallesch & Horn 1990), and visual processing, including the visual guidance of movement (Bloedel 1992; Stein & Glickstein 1992).

The cerebellum is tonically active, and presumably exerts a tonic and stabilizing influence on motor function (Llinas, 1981). Moreover, by altering its activity (e.g., Bloedel, et al., 1985; Thach, 1978), it can apparently coordinate, smooth, fine tune, as well as exert a timing influence on motor movements (Ivry, 1997; Llinas, 1981; Llinas & Sotelo, 1992). In fact, some cerebellar neurons become activated just thinking about making a movement (e.g., Dacety et al. 1990). Indeed, the cerebellum is associated not just with motor functioning, but classical conditioning and the learning of new motor programs (Llinas & Sotelo, 1992; Schmahmann, 1997).

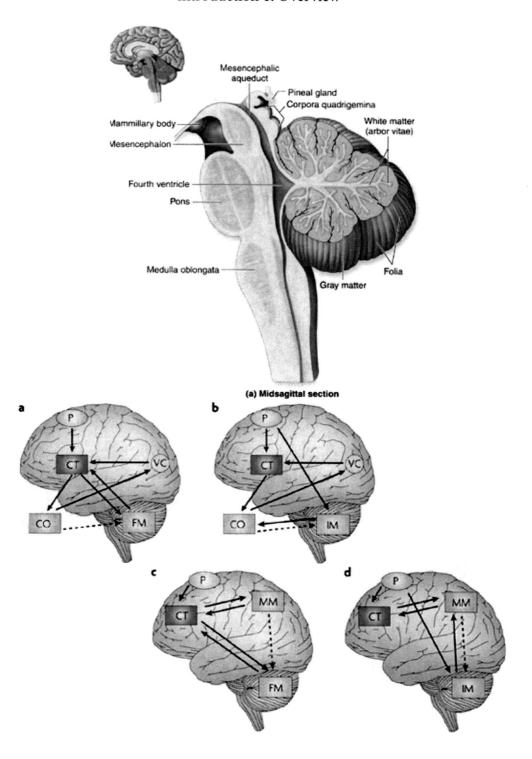

(a) Midsagittal section

For example, it has been suggested the initial acquisition of skilled movements, such as playing a guitar, requires neocortical and conscious control over motor functioning, with the cerebellum playing at best a minimal supplementary role. However, "practice makes perfect" and presumably the cerebellum immediately begins to increase its participation and slowly begins to learn the necessary movements. For example, the cerebellum becomes activated during initial learning stages (Watanabe, 1984) and as learning progress, the cerebellum may begin to acquire control over the task and may begin to associate the task and specific movements with specific changing contexts so that each context automatically triggers the movement (Thach, 1997). With time and practice, the cerebellum may slowly assume control over the associated movements, which become "automatic" and can then perform these movements with little or no help from the cerebrum which becomes free to do and think about other things.

Conversely, lesions abolish the acquisition and retention of conditioned responses (Lavond et al. 1990), and compound movements are more severely effected that simple movements. These and other findings suggests that the cerebellum may act to integrate and combine different movements, and movement sequences. Moreover, it has been proposed that climbing fibers may act to learn the task, mossy fibers may learn the "context," whereas parallel fibers integrate the context with the actual motor activity, and even correcting errors (Thach, 1997).

The Diencaphalon: Hypothalamus, Thalamus

The Diencephalon consists of the hypothalamus and thalamus.

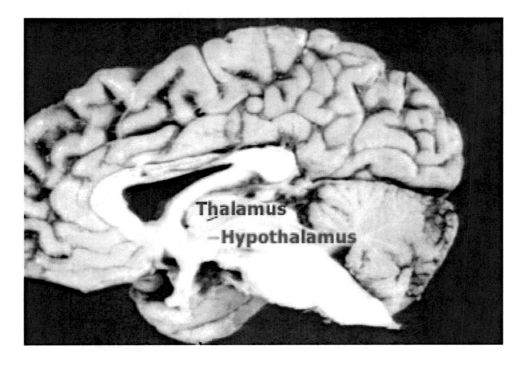

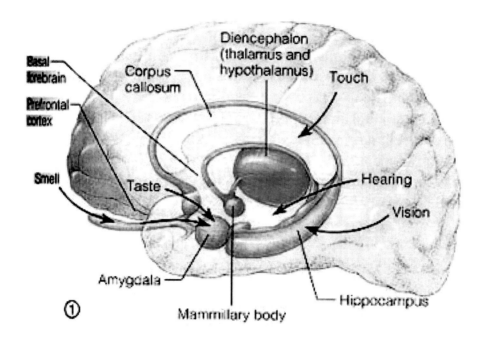

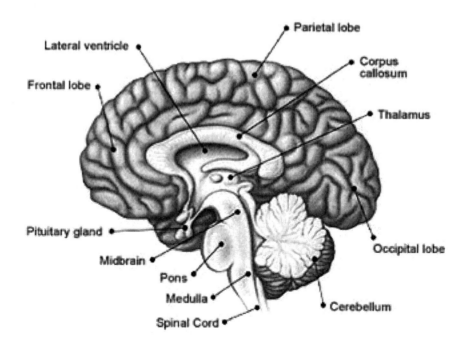

With the notable exception of olfaction (Gloor, 1997), all sensory input is first projected to the brainstem (Blessing, 1997; Vertes, 1990) and is then relayed to the immediately adjacent thalamus and hypothalamus--collectively referred to as the diencephalon ("between brain"). The diencephalon represents that rudimentary aspect of the unconscious mind that generates vague sensory impressions and diffuse emotions (Dreifuss et al. 1968; Joseph, 1992a; Olds, 1956), including pain (the thalamus), and hunger, thirst, sexual arousal, or depression and rage (the hypothalamus).

The hypothalamus could be considered the most "primitive" aspect of the limbic system, though in fact the functioning of this sexually dimorphic structure is exceedingly complex (Majdic & Tobet 2011; Orikasa & Sakuma 2010; Schoenknecht et al., 2011). The hypothalamus regulates internal homeostasis including the experience of hunger and thirst, can trigger rudimentary sexual behaviors or generate aggressive behavior or feelings of extreme rage or pleasure (Blouet et al., 2009; Lin et al., 2011; Milanski et al., 2009; Motta et al., 2009; Suzuki et al., 2010).

► **Nuclei of the Hypothalamus**

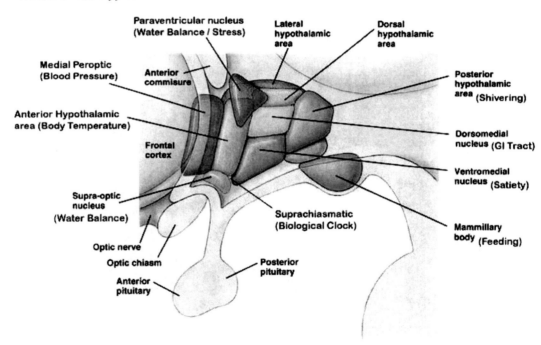

Like the brainstem, the hypothalamic portion of the diencephalon does not think or reason, but reflexively reacts--often in response to amygdala input, in which case it remain active for an extensive period of time (Dreifuss, et al., 1968; Rolls 1992). Nor are the emotions generated by this portion of the brain well differentiated. The hypothalamus may feel pleasure in general, or depression in general, or enraged in general with no differentiation, specificity, or concern for consequences other than the satisfaction of internal needs.

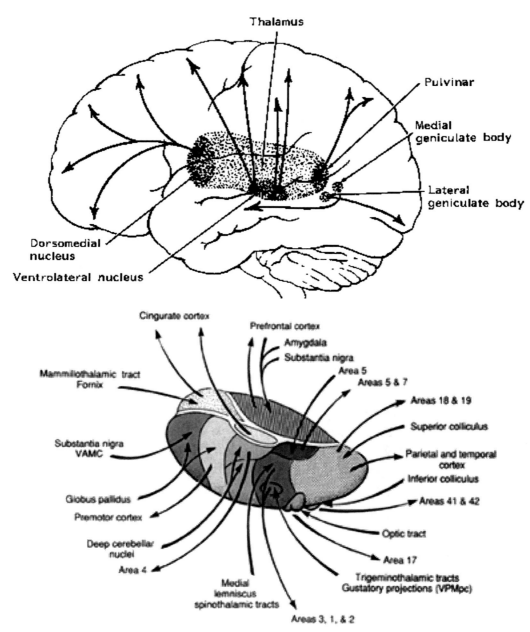

By contrast, considerable information and cognitive pre-processing occurs within the various subdivisions of the thalamus, such as the the lateral and medial geniculate nucleus (LGN & MGN), the pulvinar, the motor and subthalamus (M/ST), and the dorsal medial nucleus (DMN). These structures play a significant role in processing auditory (MGM) and visual input (LGN), the guidance of motor functions (M/ST), and the regulation of attention and arousal (DMN). For example, the LGN receives visual input directly from the retina, and then transfers this information to the visual cortex (Casagrande & Joseph, 1978). The MGN receives auditory input from the midbrain and transfer this information to the auditory cortex (Amaral et al., 1983; Pandya & Yeterian, 1985). The dorsal medial and reticular thalamus exert regulatory influences over neocortical, striatal, and limbic system arousal and influence memory and attentional functioning (Joseph, 1999a; Skinner & Yingling, 1977; Yingling & Skinner, 1977). .

In addition, the subthalamus, which is intimately associated with the amygdala, the motor thalamus and the striatum, participates in the organization and expression of gross purposeful affective-motoric behaviors (Crossman, Sambrook, & Mitchell, 1987; Parent & Hazrati, 1995). Thus the subthalamus (and striatum with which it is also intimately associated) can trigger running, kicking, punching, flailing, and a variety of oral and emotional facial expressions, or conversely "freezing" in reaction to extreme fear. The striatum and subthalamus act as an emotional-motor interface which enables humans (and other animals) to express their emotions through body language and facial expression.

Because of their role in guiding and controlling motor activities, if the subthalamus or motor thalamus is injured, patients may demonstrate a variety of hypo- or hyperactive motor abnormalities including rigidity, catatonia, and catalepsy, or conversely tremor and uncontrolled ballistic movements such as kicking, flailing, and so on (Crossman, et al., 1987; Parent & Hazrati, 1995; Royce, 1987).

Although capable of experiencing pain, the mental and perceptual functioning of the thalamus occurs outside of conscious awareness. Rather, the functioning of these various thalamic subdivisions occurs in a mental realm best described as the "preconscious;" acting to provide the conscious mind with its sensory and perceptual contents by relaying data from the brainstem to the neocortex as well as to the limbic system.

The Limbic System: Emotion & Motivation

Amygdala, Hippocampus, Septal Nuclei, Cingulate, Hypothalamus.

The principle structures of the limbic system include the amygdala, hippocampus, septal nuclei, anterior cingulate gyrus, as well as the hypothalamus. These structures represent a truly interactive system and are intimately interconnected by a number of interactive pathways, e.g. the stria terminalis, medial forebrain bundle, fornix-fimbria, amygdalofugal, and so on.

Collectively, the limbic system subserves all aspects of emotional, social, motivational and sexual functioning (Blouet et al., 2009; Gloor, 1997;

Halgren, 1992; Leutmezer et al., 1999; MacLean, 1990; Suzuki et al., 2010), as well as learning and memory (Eichenbaum et al. 1994; Mishkin, 1990; Nunn et al., 1999; Squire, 1992), and homeostatic, endocrine, and hormonal activities, including the stress response (Bao & Swaab 2010; Foley & Kirschbaum 2010; Kudielka & Wust, 2010), and even the craving for pleasure-inducing drugs (Childress, et al., 1999). With the exception of the hippocampus, electrical stimulation of each of these structures has induced feelings of extreme pleasure, and extremely negative emotions, such as fear, anger, and rage (Blouet et al., 2009; Gloor, 1997; Lin et al., 2011; Milanski et al., 2009; Motta et al., 2009; Suzuki et al., 2010). Activation of these structures, the amygdala and hypothalamus in particular, have also induced fighting, fleeing, and sexual behavior--affective motor actions made possible through the basal ganglia (e.g., corpus and limbic striatum) and the brainstem; structures which are partly under the control of the amygdala.

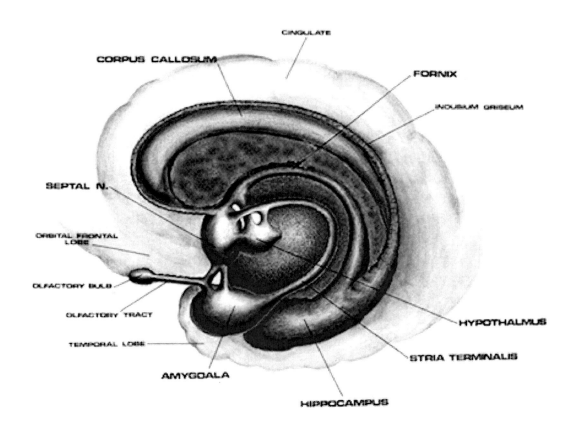

The limbic system, the amygdala and cingulate in particular, also play significant and important roles in the evolution, development and expression of language and emotional speech (Joseph, 1982, 1992a, 1999b; Newman 2010). It is these early maturing structures which provide the neurological foundations (that is, in association with the periaqueductal gray) for the development of infant speech and what has been termed, "limbic language" (Joseph, 1982, Jurgens, 1990).

Limbic System Sexuality

The limbic system is sexually differentiated such that there is a male, female, and even homosexual limbic system. In humans, sexual differentiation is initiated around 3 months after conception, and is triggered by the presence or absence of testosterone which also effects cellular development. For example, glia cells, which manufacture certain neurotransmitters and which nourish and even guide immature migrating immature neuroblasts to their terminal substrate, develop unique male-specific patterns before birth. Hence, testosterone effects neural migration and thus the organization and neural growth of the limbic system as well as the neocortex and spinal cord.

For example, the presence of fetal testosterone promotes the development of spinal motor neurons which project to the penis. Moreover, the total brain volume of the human male is about 7% larger than that of the female, and much of this differences is due to the greater volume of white matter in the male cerebrum (glia and axons), the only exceptions being the human hippocampus which is larger in the female, and the amygdala which is 16% larger in the male in total volume (Filipek, et al., 1994).

The female and male primate amygdala are sexually differentiated and have their own unique patterns of dendritic growth and organization (Nishizuka & Arai, 1981). As noted, in humans the male amygdala is 16% larger, and in male rats the medial amygdala is 65% larger than the female amygdala (Breedlove & Cooke, 1999), and the male amygdala grows or shrinks in the presence of testosterone--findings which may be related to sex differences in sexuality and aggression.

Moreover, female amygdala neurons are smaller and more numerous, and densely packed than those of the male (Bubenik & Brown, 1973; Nishizuka & Arai, 1981), and smaller, densely packed neurons fire more easily and frequently than larger ones--which may be related to the fact that females are more emotional and more easily frightened than males (chapters 7,13,15), as the amygdala is a principle structure involved in evoking feelings of fear (Davis et al., 1997; Gloor, 1997; LeDoux, 1996). Dendritic spine density in the female rat hippocampus also increases and decreases by as much as 30% during each estrus cycle (Woolley, et al., 1990) which in turn may influence memory. Indeed, in humans it has been shown that estrogen replacement therapy slows memory loss in women. In fact, it has been reported that women who take this hormone have a 54% lower chance of developing Alzheimer's disease. On the other hand, dendritic spine density can rapidly change within a few seconds (regardless of gender), as these spines can rapidly grow or disappear in response to varying experiences or lack thereof.

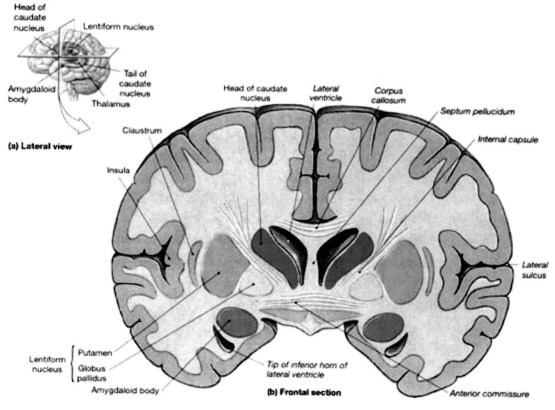

In addition, the human anterior commissure which connects the right and left amygdala/temporal lobe is sexually differentiated (Allen et al. 1989), as is primate/mammalian hypothalamus (Bleier et al. 1982; Dorner, 1976; Gorski et al. 1978; Rainbow et al. 1982; Raisman & Field, 1971, 1973)--with which the amygdala is intimately interconnected. That is, the anterior commissure is thicker in women which, coupled with her more densely packed amygdala neurons (Bubenik & Brown, 1973; Nishizuka & Arai, 1981) may account for her greater social-emotional sensitivity.

Thus, different structures of the limbic system have sex specific patterns of neuronal and dendritic organization and perform different functions depending on if one is a man or a woman.

For example, chemical and electrical stimulation of the sexually dimorphic preoptic and ventromedial hypothalamic nuclei triggers and/or increases sexual behavior in males and females (with each taking their respective sexual positions), and significantly increases the frequency of erections, copulations and ejaculations, as well as pelvic thrusting followed by an explosive discharge of semen even in the absence of a mate (Hart et al., 1985; Lisk, 1967, 1971; Maclean, 1973). In female primates, activation of these areas can also trigger maternal behavior (Numan, 1985). Conversely, lesions to the preoptic and posterior hypothalamus eliminates male sexual behavior and results in gonadal atrophy.

Likewise, activation of the sexually dimorphic amygdala--which is larger in males (Filipek, et al., 1994)-- can produce penile erection and clitoral engorgement (Kling and Brothers, 1992; MacLean, 1990; Robinson and Mishkin, 1968; Stoffels et al., 1980), and trigger sexual feelings (Bancaud et al., 1970; Remillard et al., 1983), extreme pleasure (Olds and Forbes, 1981), memories of sexual intercourse (Gloor, 1986), as well as ovulation, uterine contractions, lactogenetic responses, and orgasm (Backman and Rossel, 1984; Currier, Little, Suess and Andy, 1971; Freemon and Nevis,1969; Warneke, 1976; Remillard et al., 1983; Shealy and Peel, 1957).

Moreover, these sexually dimorphic structures also play different roles among females depending on if a woman is sexually receptive, pregnant, or lactating. For example, in a lactating female, the sexually dimorphic supraoptic and paraventricular hypothalamic nuclei (which projects to the posterior lobe of the pituitary) may trigger the secretion of oxytocin--a chemical which can trigger uterine contractions as well as milk production and which makes nursing a pleasurable experience. In fact, dendritic spine density of ventromedial hypothalamic neurons varies across the estrus cycle (Frankfurt et al., 1990) and thus presumably during pregnancy and while nursing.

Hence, the core of our personal and emotional being, the limbic system, is sexually differentiated. There is a male vs a female limbic system, and even a "homosexual" limbic system (Levay, 1991; Swaab, 1990); structures which are organized in unique sex specific dendritic and neuronal patterns and which govern sex-specific behaviors. Coupled with evolutionary (Joseph, 1999e) and early environmental influences (Joseph, 1979; Joseph & Gallagher, 1980), the sex differences in these and other structures account for many of the stereotypical sex differences in thinking, sexual orientation, aggression, and cognitive functioning (Barnett & Meck, 1990; Beatty, 1992; Dawson et al. 1975; Harris, 1978; Joseph, et al. 1978; Stewart et al. 1975) which characterized the mind of woman and man, including their sexual behaviors.

Social Behavior & The Limbic System

The amygdala are involved in all aspects of social and emotional functioning. If the right and left amygdala are destroyed, the neocortex will be denied all related social-emotional and affective input and the patient will no longer be able to recognize or feel affection for family, friends or loved one's (Lilly et al., 1983; Marlowe et al., 1975; Terzian & Ore, 1955). Although the ability to speak, think, reason, and read and write is preserved, the personal-affective contents of consciousness will have been erased.

Humans and animals subject to bilateral amygdala destruction avoid all contact with others, preferring to sit alone in isolation, and will withdraw if approached. Similarly, primates who are subject to bilateral amygdala removal lose all interest in social activity and persistently attempt to avoid contact with others. If approached they withdraw. If followed they flee. Even maternal behavior is severely affected following bilateral amygdala destruction. According to Kling (1972) mothers will bite off fingers or toes, break arms or legs, and behave as if their "infant were a strange object to be mouthed, bitten and tossed around as though it were a rubber ball".

Amygdala, Hippocampus & Memory

Although the hippocampus is not associated with emotion per se, stimulation of this structure and/or the amgydala, can trigger recent and even long forgotten memories, especially those that the patient feels to be exceedingly emotionally meaningful or personally profound, such as traumas or the recollection of the first time they had sexual intercourse (Gloor, 1997; Halgren, 1992; Penfield & Perot, 1963). The amygdala is also directly connected to the hippocampus, with which it interacts in regard to memory (Lang et al., 2009; Roozendaal et al., 2009).

The hippocampus is unique in that unlike the amygdala and other structures, almost all of its input from the neocortex is relayed via the overlying entorhinal cortex--a five layered mesocortex. As is well known, the hippocampus is exceedingly important in memory (Chadwick et al., 2010; Warburton & Brown, 2009), acting to place various short-term memories into long-term storage Katche et al., 2010; Restivo et al., (2009).

Presumably the hippocampus encodes new information during the storage and consolidation (long-term storage) phase, and assists in the gating of afferent streams of information destined for the neocortex by filtering or suppressing irrelevant sense data which may interfere with memory consolidation. The hippocampus, in conjunction with the amygdala has been implicated in the regulation of neocortical arousal, and long term memory storage and recall including the ability to remember words, conversations, and visualize one's self and the surrounding environment.

The hippocampus and amygdala displays synaptic plasticity and dendritic proliferation, and will grow additional dendritic spines in response to new learning (Engbert & Bonhoeffer, 1999), and interacts with the amygdala in the storage of the cognitive and emotional attributes of memory (Gloor, 1997; Halgren, 1992), including dreaming and what has been referred to as the primary process (Joseph, 1992a). As noted above, dendritic spines can grow and change position in response to new experiences or lack therefore, thus forming innumerable new synapses and creating vast neural networks supporting complex memories.

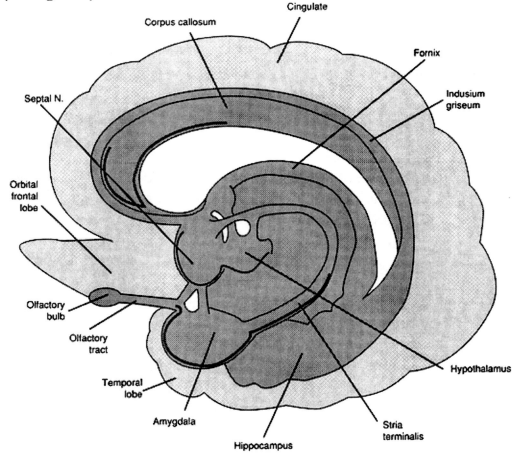

Long Term Potentiation & Memory

The growth of dendrites which sprout dendritic spines is associated with new learning and memory (Lang et al., 2004; Less et al., 2009; Restivo et al., 2009), and an excitatory phenomenon know as long term potentiation (Morris 2003; Enoki et al., 2009; Less et al, 2009; Okada et al., 2009; Whitlock et al., 2006). long term potentiation (LTP) (LTP) is associated with the growth of dendritic spines, but also occurs presynaptically (Zakharenko et al 2001; Bayazitov et al. 2007) as well as postsynaptically. LTP could be likened to a form of reverberatory activity, such that the synapse carrying the information which is to be preserved in memory, remains active for varying lengths of time.

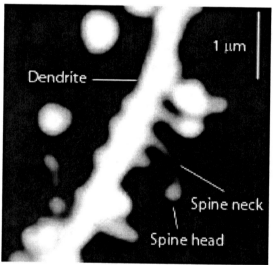

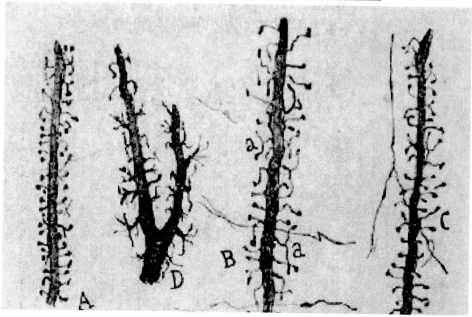

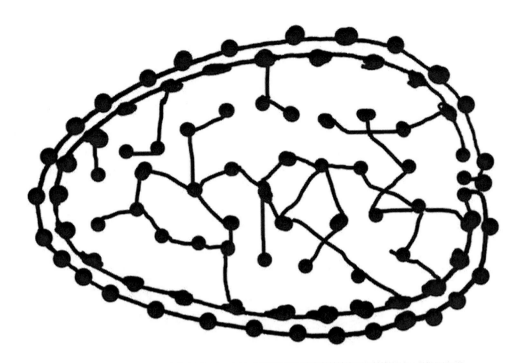

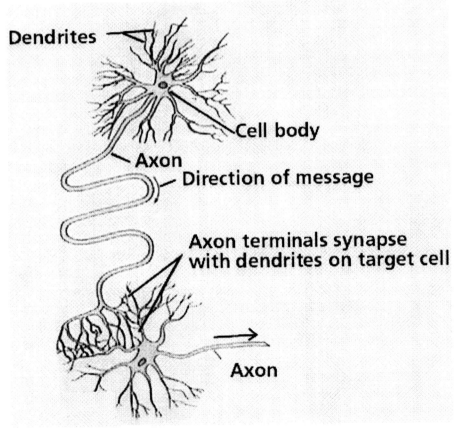

Dendrites

Cell body

Axon

Direction of message

Axon terminals synapse with dendrites on target cell

Axon

Through correlated activity both pre and post synaptically, and with the growth of dendritic spines, a new memory can be added to a neural network and/or the circuit can be modified so as to accommodate new learning (Joseph 1982). Moreover, a new set of activated neurons can be added to the network so long as their activity and interconnections become linked to the previously established pattern of excitation associated with a specific neural network. That is, certain forms of activity involving specific neurons, may represent a specific memory, and new memories may be added to this circuit when other neurons develop similar patterns of activity which then bind them in a circuit of experience. Behaviorally this may be expressed as associative learning and even classical conditioning.

LTP and the growth of dendritic spines is particularly evident in the hippocampus (Enoki et al., 2009; Hardt et al, 2010; Lang et al., 2004; Less et al., 2009; Romcy-Pereira et al., 2009; Whitlock et al., 2006). LTP is a primary feature of hippocampal neuronal activity and the hippocampus is largely responsible for the conversion of short-term memories into long term memories with different regions of the hippocampus playing different roles (Chadwick et al. 2010; Gartner & Frantz 2010) . For example, it is believed that the anterior hippocampus may prepare short-term memories for consolidation by the posterior hippocampus. The hippocampus plays a major role in long term memory, and if damage, various aspects of memory and new learning can be abolished.

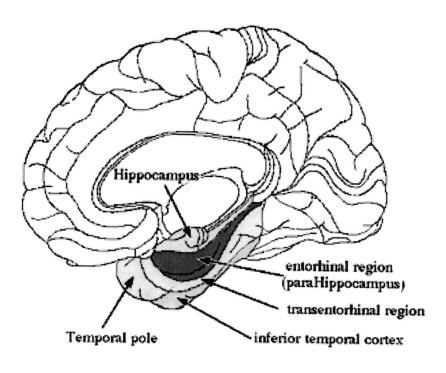

Hippocampus, Memory & Amnesia

Amnesia & H.M. Bilateral destruction of the anterior hippocampus results in striking and profound disturbances involving memory and new learning (i.e. anterograde amnesia). For example, one such individual who underwent bilateral destruction of this nuclei (H.M.), was subsequently found to have almost completely lost the ability to recall anything experienced after surgery. If you introduced yourself to him, left the room, and then returned a few minutes later he would have no recall of having met or spoken to you. Dr. Brenda Milner worked with H.M. for almost 20 years and yet she remained an utter stranger to him.

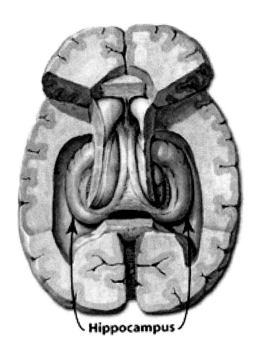

Hippocampus

H.M. was in fact so amnesic for everything that has occurred since his surgery (although memory for events prior to his surgery is comparatively exceedingly well preserved), that every time he rediscovered that his favorite uncle died (actually a few years before his surgery) he suffered the same grief as if he had just been informed for the first time.

Although without memory for new (non-motor) information, has adequate intelligence, he was painfully aware of his deficit and constantly apologized for his problem. "Right now, I'm wondering" he once said, "Have I done or said anything amiss?" You see, at this moment everything looks clear to me, but what happened just before? That's what worries me. It's like waking from a dream. I just don't remember...Every day is alone in itself, whatever enjoyment I've had, and whatever sorrow I've had...I just don't remember" (Blakemore, 1977, p.96).

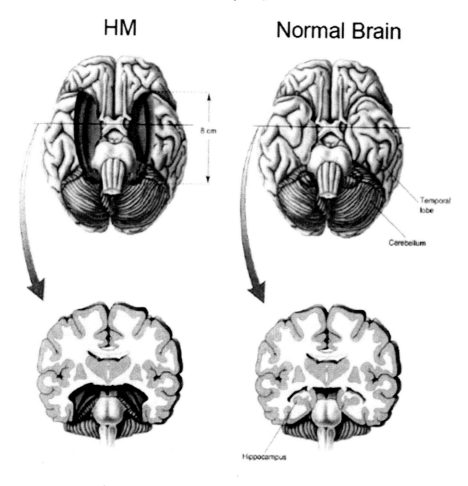

Presumably the hippocampus acts to protect memory and the encoding of new information during the storage and consolidation phase via the gating of afferent streams of information and the filtering/exclusion (or dampening) of irrelevant and interfering stimuli. When the hippocampus is damaged there results input overload, the neuroaxis is overwhelmed by neural noise, and the consolidation phase of memory is disrupted such that relevant information is not properly stored or even attended to. Consequently, the ability to form associations (e.g. between stimulus and response) or to alter preexisting schemas (such as occurs during learning) is attenuated.

The Cingulate & Entorhinal Cortex

The different structures of the limbic system subserve different as well as overlapping functions ranging from primitive emotions (e.g., hypothalamus) to the spiritually profound (e.g., the amygdala). From an evolutionary perspective, some structures are exceedingly ancient and have a pedigree extending almost a half billion years backwards in time, e.g. the amygdala/striatal/hippocampus, hypothalamus, brainstem.

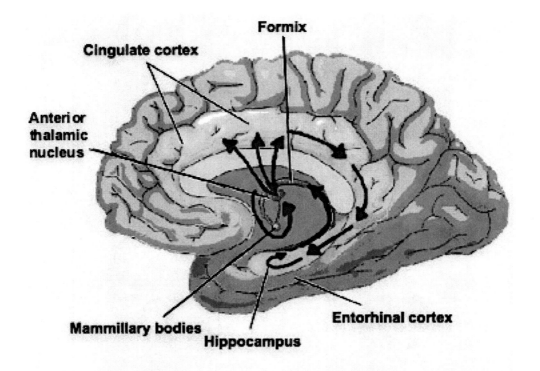

Others of more recent vintage, such as the cingulate gyrus, may have first begun to evolve around 200 million years ago (MacLean, 1990). This more recent evolutionary origin is also reflected by those functional capacities it mediates, including complex cognitive-affective activities such as maternal-infant behavior and emotional speech.

That the cingulate has evolved more recently is also evident structurally. For example, with the exception of the cingulate, the structures of the limbic system are comprised of allocortex. Allocortex has three layers with pyramidal cells sandwiched between layers I and III. The cingulate consists of mesocortex (also referred to as "paleocortex" and "transitional cortex"), which consists of five layers.

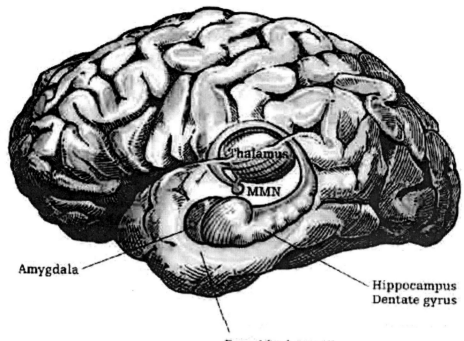

Amygdala

Hippocampus
Dentate gyrus

Entorhinal cortex
Perirhinal and parahippocampal cortices

Although the entorhinal cortex --the "gateway to the hippocampus-- may have begun to evolve at the same time as the cingulate, unlike this latter structure, the entorhinal cortex appears to have continued to "evolve" and add new layers. The entorhinal cortex, which receives and relays information to and from the neocortex and hippocampus, consists of between 7 and 8 layers (Braak & Braak, 1992; Ramon y Cajal, 1902/1955; Rose, 1926). This seven to eight layer organization may well partly explain the unique importance of the entorhinal cortex in complex cognitive processing, for it receives input from the hippocampus and all neocortical association areas information which it apparently integrates, and in conjunction with the amygdala and in particular the hippocampus (which it partly coats), stores in memory (Gloor, 1997; Squire, 1992). The entrohinal cortex, which partly surrounds the hippocampus and is interconnected with the amygdala, appears to be a supra-modal memory center.

The Limbic And Corpus Striatum

Within 50 million years of the close of the Cambrian "Explosion" (500 million years ago), cartilaginous sharks began to swim and patrol the primeval seas. Sharks are considered a "living fossil" and dissection of the shark brain reveals a brainstem, diencephalon, and a forebrain consisting of a dorsal/ventral amygdala-striatal mass with a primordial hippocampal-striatum centered at its dorsal core. Together the amygdala, striatum, and hippocampus, formed the forebrain, the dorsal pallium (Gloor, 1997; Haberly 1990; Herrick, 1925; Stephan & Andy, 1977; Ulinksi, 1990).

Although the recipient of visual input, transferred from the brainstem, the primordial amydala-striatal-hippocampus, was dominated by the olfactory lobe--a dominance which was exaggerated further when animals left the sea and began to wonder upon dry shores. With the evolution of amphibians and reptiles, the forebrain expanded and the amygdala, striatum, and hippocampus began to differentiate, and were pushed apart (Gloor, 1997; Haberly 1990; Herrick, 1925; Nieuwenhuys & Meek, 1990ab; Smeet, 1990; Stephan & Andy, 1977; Ulinksi, 1990). Nevertheless, the amygdala remains intimately associated with the striatum, which in turn responds to amygdala concerns.

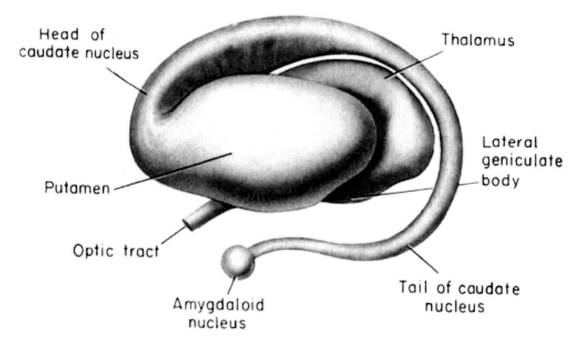

In the human brain, the striatum is located anterior to the thalamus and is a major component of the basal ganglia. Considered broadly, the basal ganglia consists of the subthalamic nucleus, portions of the midbrain, the limbic striatum, the amygdala, and the globus pallidus and putamen (the lenticular nucleus) and the caudate nucleus. The putamen and the caudate are referred to as the corpus striatum, with the head of the caudate extending deep into the frontal lobe and its tail merging with the amygdala. Beneath the corpus striatum is the limbic striatum (or extended amygdala) which is comprised of the substantia innominata, the nucleus accumbens, and the olfactory tubercle.

The limbic and corpus striatum serve, in part, as the motor-component of the limbic system, and receive input directly from the amygdala. Hence, the striatum reacts to certain types of visual and olfactory stimuli which are deemed by the amygdala to be emotionally significant, with gross motor movements, e.g., kicking, hitting, running, flailing.

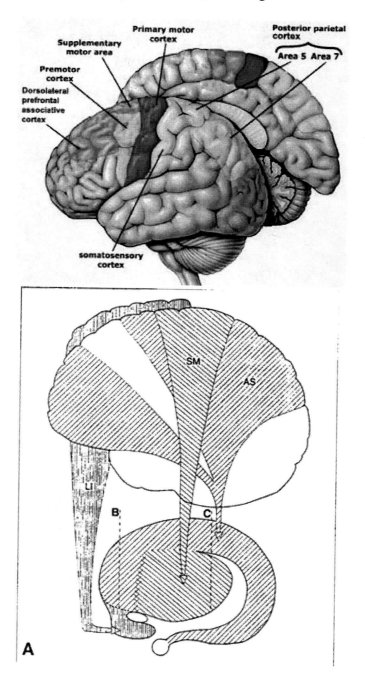

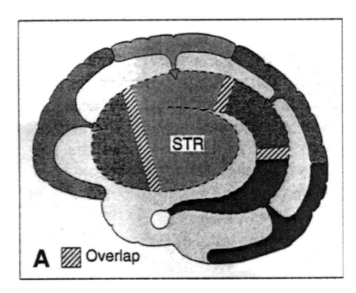

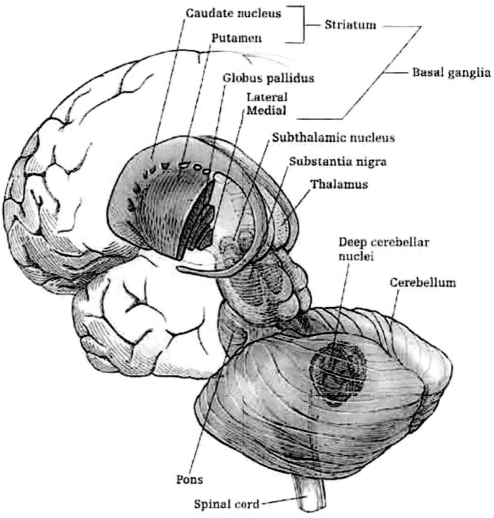

Because of its importance in movement, if the corpus striatum were injured, the patient may become stiff and rigid, such as occurs with Parkinson's disease, or they may involuntarily twitch, jerk, kick, and flail about with their limbs--depending on which aspects of the basal ganglia are injured. In fact, a host of cognitive, affective, and motor disturbances may ensue, including Parkinson's and Alzheimer's diseases, chorea, hemiballismus (uncontrolled kicking, punching), restless leg syndrome, catatonia, schizophrenia, obsessive-compulsions, or depression (e.g. Aylward et al. 1994; Castellanos et al. 1994; Chakos et al. 1994; Deicken et al. 2005; Ellison 1994; Harrison, 2009; Perez-Costa 2010; Turjanski et al, 2009; Welter et al. 2011). Patients with abnormalities and reduced functional activity of the striatum may develop obsessive compulsive motor movements, often involving the hands, or increasingly lose inhibitory control over motor functioning such that may display signs of mania. In fact, children with hyperactivity and impulse control disorder, have been shown to display reduced striatal activity, which is increased when treated with Ritalin. However, in "normal" children, Ritalin decreases striatal activity.

Although the dorsal and ventral aspects of the striatum play roles in motor functioning, the limbic striatum is more concerned with emotion and memory. Hence, damage to this structure is associated with loss of memory including Alzheimer's disease.

The Limbic System Vs Neocortex: Consciousness

The contents of consciousness, that is, the contents of the neocortex, are initially derived from the limbic system and thalamus, which provide sensory, perceptual, and emotional input. Structures such as the amygdala and thalamus project to almost every region of the neocortex, and if denied limbic and thalamic input, consciousness would be extinguished and the ability to become conscious of the external or internal world would be denied.

However, although "conscious" the limbic system does not appear capable of self-consciousness or self-reflection. The brain of the shark, amphibian, and reptile, consist of limbic and brainstem tissue, and there is no evidence of self-consciousness, or thinking or thought among these creatures. Rather, they reflexively react. Of course, the limbic system has also continued to evolve with the evolution of mammals and then humans--and in part, the neocortex is the consequence of the evolutionary expansion.

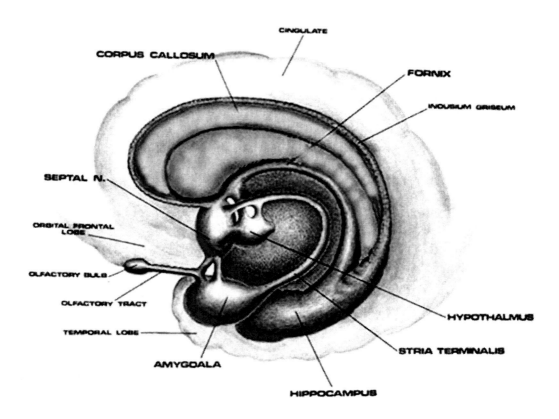

Hence, whereas the neocortical aspects of the conscious human mind are concerned with the more rational and linguistic aspects of experience, the limbic aspects of the human mind are associated with the emotional and even the hallucinatory aspects of experience, including those features associated with what has been described as the primary process. By contrast, the brainstem, controlling waking and sleeping, as well as the rhythmic aspects of vegetative functioning, could be identified with the most primitive regions of the psyche as its functional activity is for the most part, completely beyond conscious scrutiny or control.

The Neocortex (Gray Matter)

Perception, cognition, fine-motor expression, and computational processing, is made possible by neurons, the majority of which are pyramidal neurons. Most pyramidal neurons are located in the neocortical mantle of the lobes of the brain, which gives this outer coating its grayish appearance. Over 90% of the gray matter is located in the neocortex. The neocortex ("new cortex") likely first began to evolve between 100 million to 150 million years ago (MacLean, 1990).

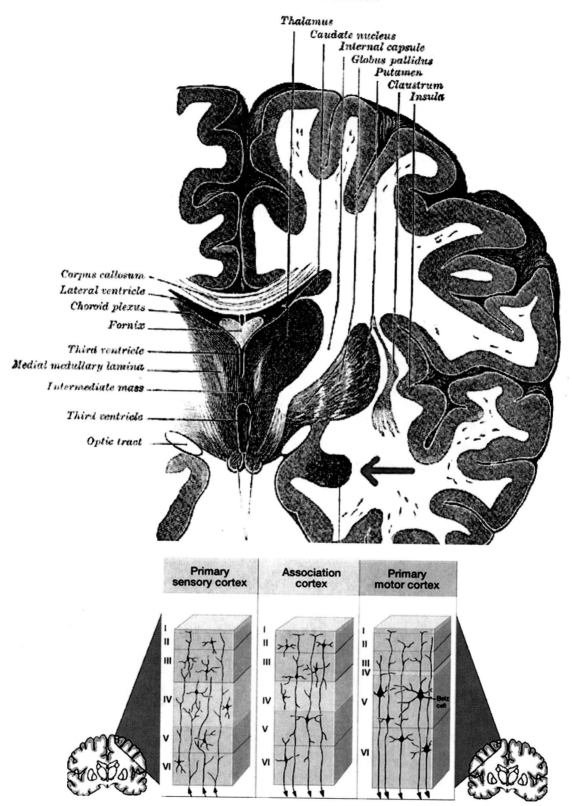

Those neurons which project to neurons in the next column or to those neurons in an upper or lower layer, are referred to as local circuit neurons. Those which project from the neocortex to the brainstem, or from one half of the brain to the other, are referred to as long distance neurons (Peters & Jones, 1984).

Neocortical Layers

Classically, the neocortex is said to consist of six to seven layers when in fact it consists of numerous layers which vary depending on brain area (Braak & Braak, 1992; Peters & Jones, 1984; Ramon y Cajal, 1902/1955; Rose, 1926). For example, the deepest layer, neocortical layer VI, consists of two distinct layers (VIa and VIb). In the occipital lobe, three additional layers (i.e. sublayers) can be distinguished within layer IV (which also receives considerable thalamic input and is very thick). By contrast, within the motor areas of the frontal lobe, layer IV is exceedingly thin (as there is comparatively minimal thalamic input), whereas layer V is exceedingly thick, It is layer V of the frontal motor areas which contribute the bulk of axons that form the descending corticbulbar, corticopontine, and corticorubral brainstem pathways which establish contact with cranial nerve and sensory and spinal motor neurons (Brodal, 1981; Kuypers & Catsman-Berrevoets, 1984). Likewise in the temporal neocortex layer V is relatively thick as are layers I and VI (since much of the temporal lobe is association and assimilation cortex). As noted, the entorhinal cortex, the "gate way to the hippocampus" and which is located along the medial surface of the temporal lobe, consists of between 7 and 8 layers (Braak & Braak, 1992; Ramon y Cajal, 1902/1955; Rose, 1926).

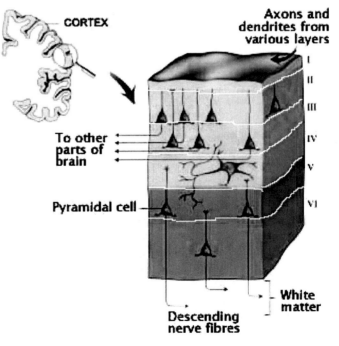

CEREBRAL CORTEX

cortical surface

I

II

III

IV

V

VI

white matter

PYRAMIDAL NEURON

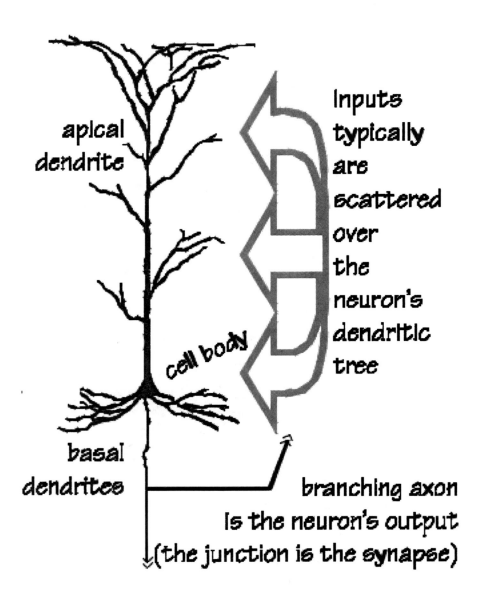

apical
dendrite

cell body

basal
dendrites

Inputs
typically
are
scattered
over
the
neuron's
dendritic
tree

branching axon
is the neuron's output
(the junction is the synapse)

Hence, the thickness, layering, and composition of the human neocortex varies from lobe to lobe and actually consists of from 7 to 9 (or more) layers rather than 6.

Specifically, layer I is referred to as the Molecular Layer and consists of Golgi II cells and horizontal cells. Layer I receives innumerable dendrites from local circuit neurons located in the lower layers. Layer I, however, actually contains few neurons and is mostly made up of tangentially running axons and horizontally running birfucating apical dendrites received from the pyramidal cells of the lower layers (Peters & Jones, 1984). Layer II is referred to as the External Granular Layer, and consists of densely packed small pyramidal, stellate, and granule cells. Most of the neurons in layer II are local circuit neurons which project to adjacent columns and adjacent layers.

CEREBRAL CORTEX

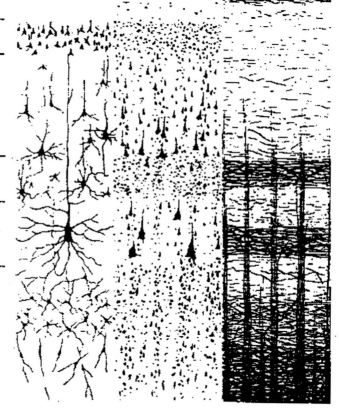

1. MOLECULAR LAYER

2. EXTERNAL GRANULAR LAYER

3. EXTERNAL PYRAMIDAL LAYER

4. INTERNAL GRANULAR LAYER

5. INTERNAL PYRAMIDAL LAYER

6. MULTIFORM LAYER

Layer III is the Pyramidal Layer and consists of medium pyramidal cells which project axons to distant areas of the neocortex. Hence, the neurons of layer III can be considered long distance neurons.

Layer IV, the Internal Granular Layer, has a granular appearance and consists of small pyramidal, granule, and stellate (starshaped) cells and receives massive axonal projections from the thalamus. These neurons are predominantly local circuit, and project to adjacent columns and layers. That is, upon receiving and analyzing thalamic input, the neurons of layer

IV transfer this data to adjacent neurons for additional analysis. Because the primary, secondary and association sensory areas receive considerable thalamic input, layer IV is relatively thick--except in the motor cortex.

Layer V is the Ganglionic Layer and consists of large and medium size pyramidal cells, including, in primary motor cortex (Brodmann's area 4) the giant cells of Betz. The pyramidal neurons of layer V are long distance neurons, and give rise to descending axons which form the corticospinal, pyramidal, corticobulbar, corticopontine, and corticorubral brainstem pathways which establish contact with cranial nerve and sensory and spinal motor neurons (Brodal, 1981; Kuypers & Catsman-Berrevoets, 1984). It is these "pyramidal" and cortico-spinal neurons which make purposeful, fine motor movement possible. Approximately 31% of the corticospinal tract arises from the pyramidal cells located in the primary motor areas 4, with the remainder arising from the frontal motor associations areas 6, 8, and the primary somesthetic areas 3,1,2, with a scattering of fibers being contributed by the occipital and temporal lobe, as well as limbic system structures.

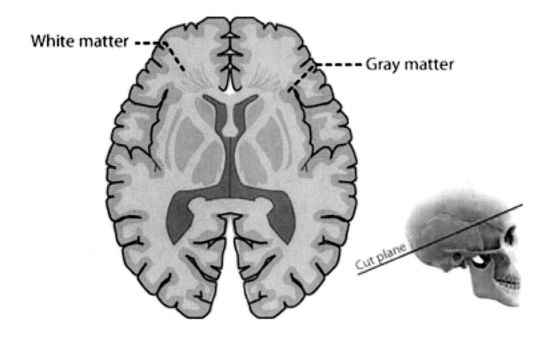

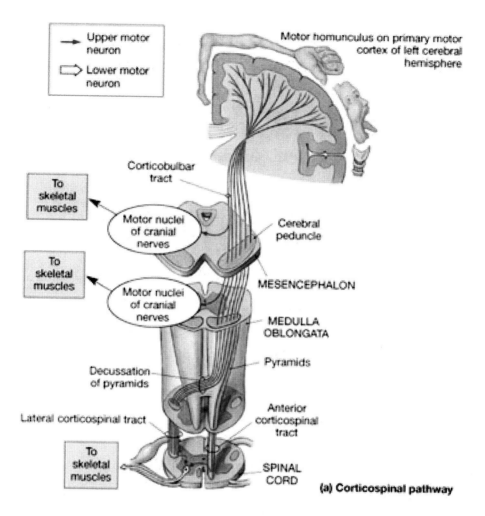

Upper motor neuron

Lower motor neuron

Motor homunculus on primary motor cortex of left cerebral hemisphere

Corticobulbar tract

To skeletal muscles

Motor nuclei of cranial nerves

Cerebral peduncle

To skeletal muscles

Motor nuclei of cranial nerves

MESENCEPHALON

MEDULLA OBLONGATA

Pyramids

Decussation of pyramids

Lateral corticospinal tract

Anterior corticospinal tract

To skeletal muscles

SPINAL CORD

(a) Corticospinal pathway

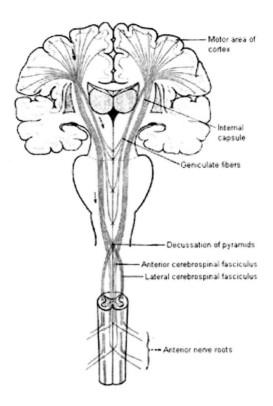

Layer VIa is the Multiform Layer and contains pyramidal, fusiform, and spindle shaped cells, whereas Layer VIb consists of predominantly of spindle shaped cells. These are predominantly local circuit neurons, and receive considerable input from the brainstem.

Cytoarchitextural, Neuronal, & Chemical Organization of the Neocortex

Korbinian Brodmann detailed the regional variation in the cytoarchitectural organization of the cortex, and conducted detailed comparative studies of numerous species, each of which displays common as well as varying patterns of cytoarchitexture and gyral folding. Based on these cytoarchitextural differences and commonalities, Brodmann divided the cortex into distinct regions and created cytoarchitextural maps of the brains of a variety of species, including humans. For examples, Brodmann's area 17 is synonymous with the primary visual cortex, whereas Brodmann's areas 3,1,2, denote and identify the primary somesthetic receiving areas.

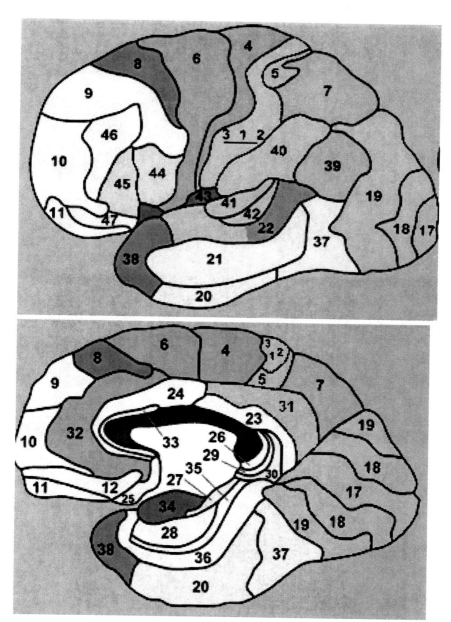

However, although these area differ in regard to organization, what they share in common is a preponderance of pyramidal cells. As noted, pyramidal cells are also the largest and are more numerous than any other neocortical neuron (Peters & Jones, 1984). Pyramidal neurons account for up to 3/4 of all neocortical cells. Pyramidal neurons also serve as both local-circuit and long-distance neurons and generally receive two types of synaptic contacts referred to as Gray types I and II which differ in synaptic morphology and (respectively) excitatory vs inhibitory influences (Peters & Jones, 1984). However, almost all pyramidal cell are excitatory and use glutamate and aspartic acid as transmitters (Tsmoto, 1990).

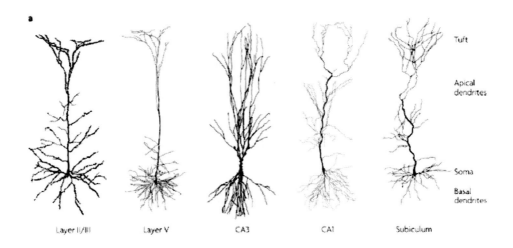

Layer II/III   Layer V   CA3   CA1   Subiculum

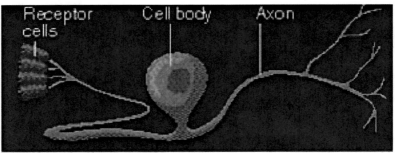

(a) Sensory Neuron

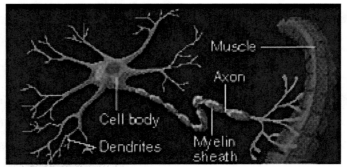

(b) Motor Neuron

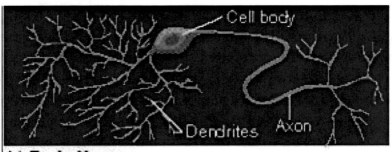

(c) Brain Neuron

Pyramidal cells can also be classified as Golgi I and II cells. However, of all neocortical neurons, only 10% are Golgi type I neurons, the main source of long-distance (excitatory) axons, the majority are interneurons, i.e. local circuit neurons which in turn provide almost 90% of cortical axons and dendrites. Approximately 95% of Golgi type I long distance axons interconnect distant regions within the same hemisphere and only about 5% cross the corpus callosum, the fiber pathways which link the right and left hemisphere (Peters & Jones, 1984).

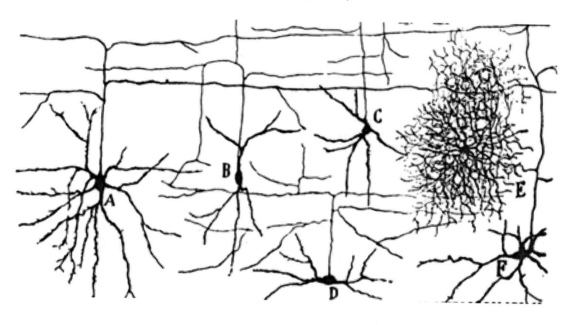

Non-pyramidal cells also function as local-circuit (inter-) neurons, and connect adjacent cells, layers, and cell columns, and display the greatest degree of morphological diversity. These include stellate cells, bipolar cells, chandelier cells, and basket cells. Many of these cells are inhibitory and may use GABA as a neurotransmitter, though the majority in fact appear to be excitory and use glutamate (Peters & Jones, 1984). Presumably, non-pyramidal, local circuit (interneurons) act to fine tune information processing via inhibitory filtering and selective excitatory transmission. They also serve to integrate and assimilate information received in adjacent regions of the neocortex.

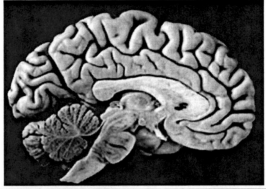

In addition to glutmate and GABA, neocortical neurons contain and respond to peptides, including substance P, corticotropin releasing factor, and opiates. The peptide containing neurons tend to congregate in layers II, III, and IV (Jones & Hendry, 1986).

Information processing throughout the brain is also dependent on glia and non-neuroglia elements. Glia serve a supportive and nurturing role, and may also act to store information. During embryonic brain development, radial glia fibers act to guide migrating neurons to the neocortex, and some glia also form myeline sheaths which surround axons, thus serving as a form of insulation which promotes information transmission.

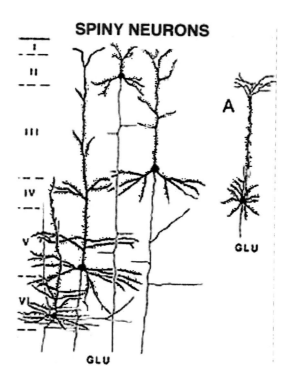

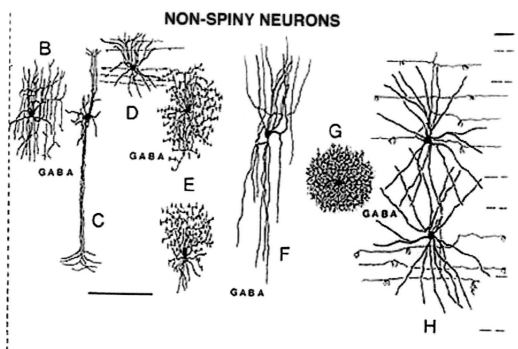

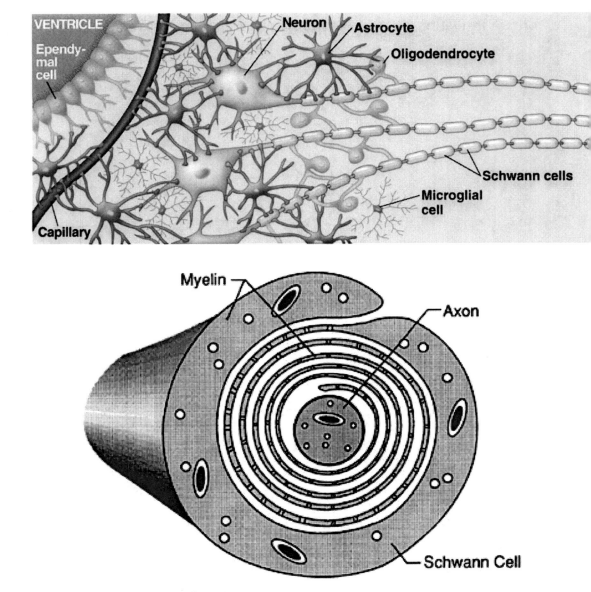

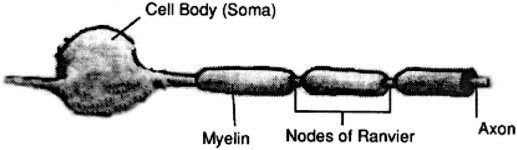

Glia and non-neuroglia elements make up almost 70% of the volume of the neocortex. Of the remainder, 22% consists of axons and dendrites, with the body (soma) of the neuron comprising only 8% (Peters & Jones, 1984).

Neocortex & The Conscious And Unconscious (Emotional) Mind

The human cerebrum can be subdivided into frontal, temporal, parietal, and occipital lobes, as well as the limbic system, striatum and diencephalon. The brainstem and cerebellum are not considered part of the cerebrum, but instead comprise the "hind brain."

Neocortex & Consciousness The cerebrum constitutes nearly 90% of the volume of the brain, and is 50 times larger than the brainstem and 8 times larger than the cerebellum (Filipek, et al., 1994). More than 60% of the cerebrum consists of gray matter, and less than 40% consists of white matter. The amygdala, basal ganglia, diencephalon and hippocampus, make up less than 3% of the central gray, whereas more than 90% of the gray matter is found within the neocortex, which reflects the dominant role the neocortex plays in virtually all fully-integrated functions of the CNS.

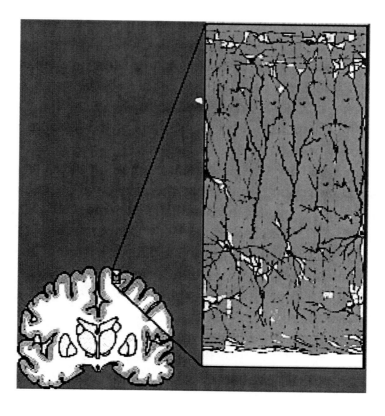

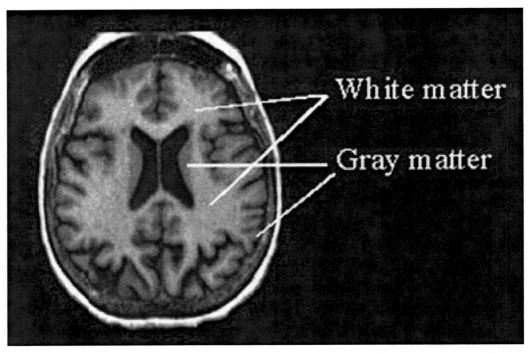

The neocortex is clearly associated with what is classically considered the conscious mind--a consciousness that knows it is conscious, and which includes the capacity to think and reason, to abstract and consciously reflect upon the self and the world, as well as the ability speak grammatically and read, write, compose, and recite the music of poetry. Consciousness, however, is modular, with different regions of the neocortex engaging in both localized as well as parallel processing.

For example, as based on functional imaging, it has been demonstrated that language processing, and silent mental activities, such as thinking, or generating inner speech, activates the neocortex of the frontal lobes (Paulesu, et al., 1993; Peterson et al., 1988; Demonet, et al., 1994). Reading and language processing also activates the neocortex of the temporal lobes (Bookheimer, et al., 1995; Bottini, et al., 1994; Fletcher et al., 1995; Shaywitz, et al., 1995; Warburton, et al., 1996) and the left inferior parietal lobule (Bookheimer, et al., 1995; Price, 1997). Thus, the IPL, Wernicke's area, and frontal lobes become active across a variety of language and non-language problem-solving and thinking tasks (Ben-Shahar et al., 2003; Lehmann et al. 2009; Szaflarski et al., 2006; Tyler & Marslen-Wilson 2008; Vigneau et al. 2006). Moreover, during language processing there is activity in the brainstem, the cerebellum, and various limbic structures, especially the mesocortical anterior cingulate gyrus.

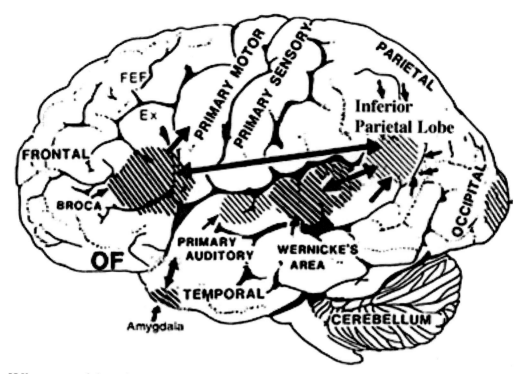

When considered very broadly, it could be argued that the neocortex is associated with the conscious, rational mind, and conscious-awareness that we identify as distinctly human--a consciousness that knows it is conscious (Joseph, 1982, 1988a). By contrast, it has been argued (Joseph, 1992ab) that the limbic system represents that aspect of the mind classically referred to as the "unconscious" (Freud, 1900) and the "collective unconscious" (Jung, 1945). And just as it has been theorized that the unconscious continually supplies the conscious mind with all manner of impulses, imagery, and ideas (e.g. Freud, 1900), it is now apparent that the limbic system does likewise, and that those aspects of consciousness associated with the neocortex are often driven by these unconscious limbic mental realms and associated impulses.

The Frontal, Parietal, Temporal And Occipital Lobes

The rational, logical, linguistic, and self-reflective aspects of consciousness, therefore, are associated with the neocortical shroud which envelops the brain. Neocortical consciousness, however, also appears to be somewhat modular with the different lobes of the brain (and the right vs left hemisphere) subserving different aspects of consciousness and perceptual and personality functioning.

As is common knowledge, the six to seven layered neocortical shroud which encompasses and envelopes the old brain, can be divided into the frontal lobes which comprise the anterior half of the human telencephalon, and the parietal, occipital and temporal lobes which are located in the posterior half of the cerebrum, each of which contributes differently to the mosaic of mind including personality.

The Frontal Lobes

The frontal lobes has been referred to as the "senior executive" of the brain and personality and is associated with goal formation, long term planning skills, the ability to consider multiple alternatives and consequences simultaneously, as well as memory search and retrieval. Because it is interlocked with the thalamus, limbic system, brainstem, and the parietal, occipital, and temporal lobes, the frontal lobes are provided multiple streams of input and are constantly informed as to the processing which takes place in other regions of the brain. Moreover, it can act to inhibit, suppress, or enhance perceptual and information processing, including learning and memory through inhibitory and excitatory influences directed to the thalamus, brainstem, or the different lobes of the brain (Joseph, 1999a).

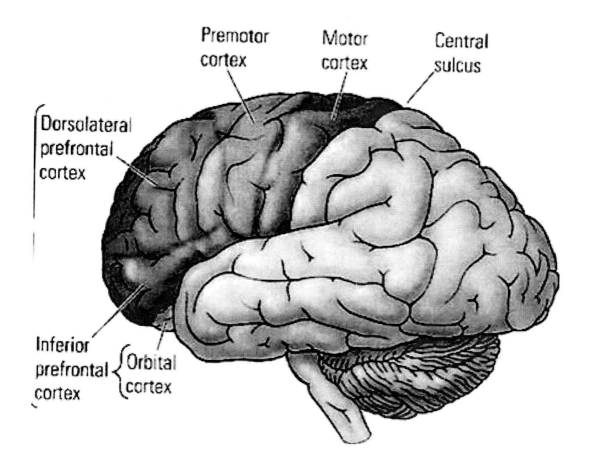

Premotor cortex   Motor cortex   Central sulcus

Dorsolateral prefrontal cortex

Inferior prefrontal cortex   Orbital cortex

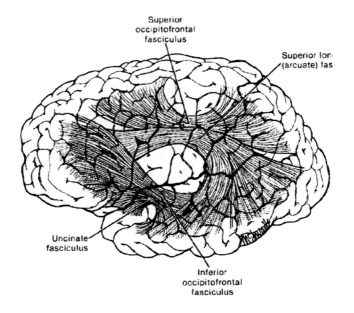

For example, the frontal lobes interact directly with the temporal lobe (which contains the hippocampus and amygdala) in order to ensure that certain percepts are attended to, committed to memory, and later remembered. As demonstrated through functional imaging, it has been determined that the ability to remember a visual or a verbal experience, is directly correlated with activation of the right or left frontal lobe and the temporal lobes during that experience (Brewer et al., 1998; Wagner et al., 1998). According to Wagner et al., (1998, p. 1190), "what makes a verbal experience memorable partially depends on the extent to which left prefrontal and medial temporal regions are engaged during the experience." The degree of this activity, therefore, can be used to predict which experiences will be "later remembered well, remembered less well, or forgotten" (Brewer et al., 1998, p. 1185). Thus the frontal lobes appear to be acting directly on and in concert with the amygdala and hippocampus in regarding to memory functioning.

In addition to influencing arousal, memory, and perceptual actiivty, the right and left frontal motor areas control fine motor activity, including the emotional and non-emotional aspects of expressive speech (Joseph, 1982, 1988a, 1999a,e; Ross, 1993). Both the right and left frontal lobes in fact become functionally active during language tasks (Peterson et al., 1988), and receive auditory-linguistic input that has been organized and transmitted from the posterior speech zones, the inferior parietal lobule, and in the left hemisphere, Wernicke's area in the superior temporal lobe. Damage to the left frontal lobe can thus result in expressive aphasia, whereas right frontal injury may disturb the melody and prosodic aspects of speech.

Because the frontal lobes serve the "senior executive" of the brain and personality, if the frontal lobes are injured, all aspects of personality functioning may become severely abnormal (Fuster, 2007; Joseph, 1986a, 1988a, 1999a; Stuss 2009). Patients may become lethargic, apathetic, or conversely, disinhibited, impulsive, aggressively sexual, and display what has been classically referred to as the frontal lobe personality.

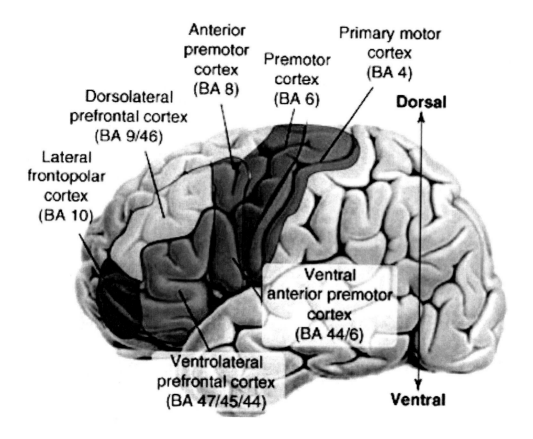

One individual who was described as quite gentle and sensitive prior to his injury, subsequently raped and brutalized several women. Similar behavior has been described following frontal lobotomy. As stated by Freeman and Watts (1943, p. 805): "Sometimes the wife has to put up with some exaggerated attention on the part of her husband, even at inconvenient times and under circumstances which she may find embarrassing. Refusal, however, has led to one savage beating that we know of, and to an additional separation or two" (p. 805). Curiously, in these situations Freeman and Watts (1943, p. 805) have suggested that "spirited physical self-defense is probably the best strategy of the woman. Her husband may have regressed to the cave-man level, and she owes it to him to be responsive at the cave-women level. It may not be agreeable at first, but she will soon find it exhilarating if unconventional."

However, since the frontal lobes serve so many functions, patients may display different symptoms depending on if the lesion impacts the right, left, orbital, or medial aspects of the frontal lobes (Joseph, 1999a). Oatients may display severe apathy, depression, schizophrenia or aphasic speech (left frontal), mania, disinhibition, and confabulation (right frontal), obsessive compulsions (orbital-striatal) or catatonia (medial frontal).

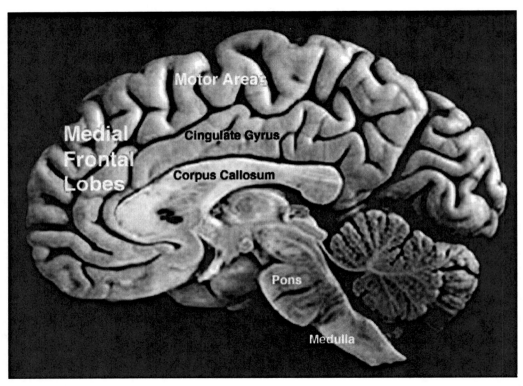

For example, a soldier who received a gunshot wound that passed completely through the frontal lobes, "layed in a catatonic-like stupor for two months, always upon one side with slightly flexed arms and legs, never changing his uncomfortable position; if he were rolled into some other position, he would quickly get back into his former one. He did not obey commands, but if food and drink were given to him, he swallowed them naturally. He was incontinent, made no complaints, gazed steadily forward and showed no interest in anything. He could not be persuaded to talk, and then suddenly he would answer quite correctly about his personal affairs and go back to mutism. From time to time he showed a peculiar explosive laugh, especially when his untidiness was mentioned". Incredibly, the patient "was eventually returned to active duty" (Freeman & Watts 1942, pp 46-47).

The Frontal Motor Areas

Movement and motor functioning are dependent on the functional integrity of the basal ganglia, brainstem, cerebellum, cranial nerve nuclei, the motor thalamus, spinal cord, as well as the primary, secondary and supplementary motor areas of the frontal lobes (Chouinard & Paus 2006; Dum & Strick 2005; Nachev et al. 2008; Verstynen, et al. 2011). These areas are all interlinked and function as an integrated system in the production of movement (Mink, 1997; Mink & Thach, 1991; Parent & Hazrati 1995; Passingham, 1997; Takara et al. 2011).

However, as to fine motor movements including those involved in the articulation of speech, these are almost completely dependent upon the functional integrity of the primary motor areas located along the precentral gyrus (area 4), and within which are neurons which represent and control the muscles of the hands, fingers, and oral laryngeal musculature.

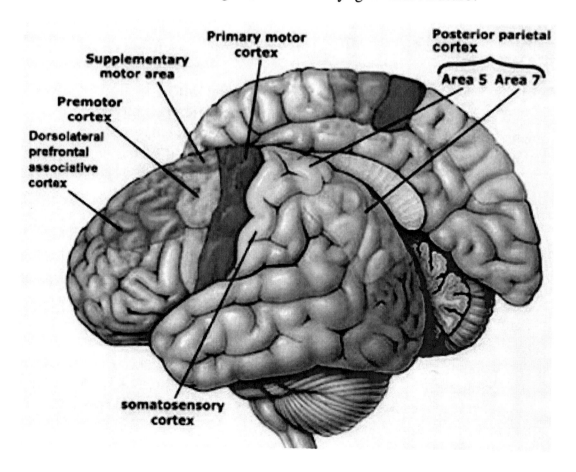

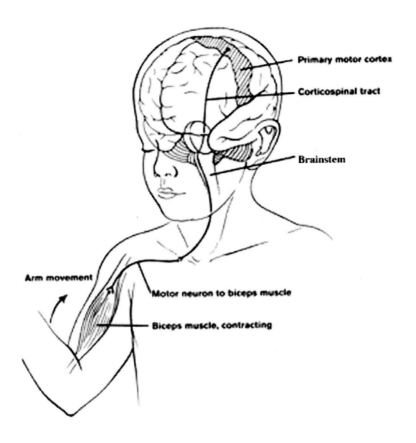

The primary motor areas are in turn dependent upon motoric impulses which are organized in premotor and the supplementary motor cortex-- the latter of which is located along the medial wall of the hemispheres. Considered from a very broad and simplistic perspective, it could be said that primary area is programmed and under the control of the secondary and supplementary motor areas as well as the "prefrontal" and other areas of the cerebrum, although neurons in the primary area also become active prior to and during movement (Passingham, 1997). For example, Exner's writing area is in part, within areas 6 and becomes active prior to (as well as during) hand movements and appear to program hand movements, whereas the frontal eye fields (within areas 6,8,9) becomes active prior to (as well as during) eye movements and appears to program eye movements. As noted above, the primary area representing the oral-laryngeal musculature, is programmed by Broca's expressive speech area areas 45, 46. Broca's area also becomes active prior to vocalization and during subvocalization as indicated by functional imaging.

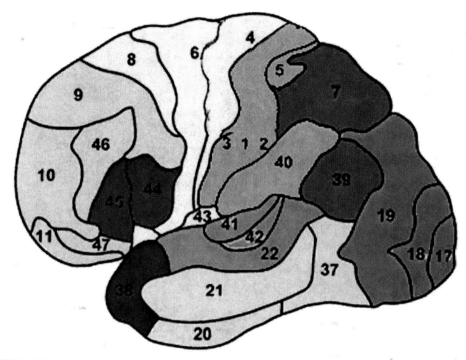

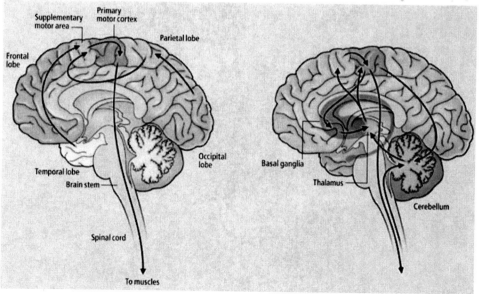

Therefore, the motor areas of frontal lobes control body movement and fine motor activity. A motor map of the body is maintained within the neocortex of the frontal lobes, and if damaged, the individual will become paralyzed on one half of the body; the extent of the paralysis depending on the extent and depth of the damage.

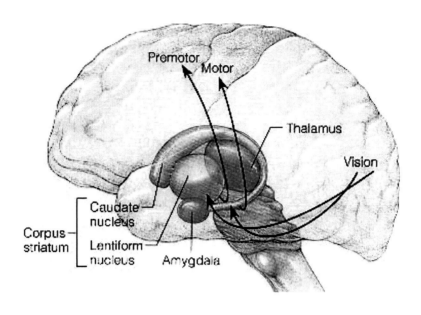

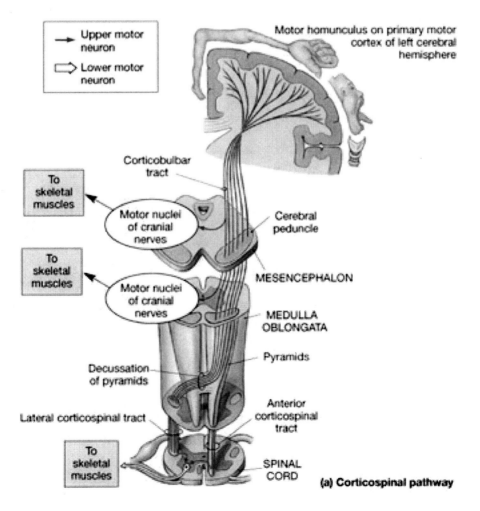

(a) Corticospinal pathway

The frontal lobes control muscular activity and movement through the pyramidal tract (also known as the cortico-spinal tract). Pyramidal neurons project to the motor centers in the striatum, thalamus, brainstem, and the spinal cord. In this way, gross and fine motor movement can be controlled by the senior executive of the brain and personality.

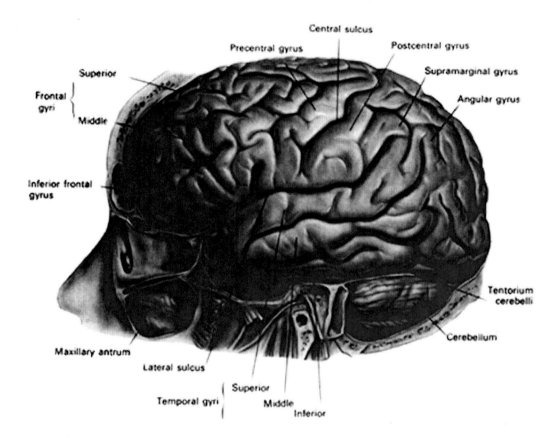

The Temporal Lobes

Whereas the frontal lobe act to regulate personality and emotion, such as through inhibitory projections it maintains with the limbic system, the temporal lobes are the source of one's personal subjective and emotional identity (Gloor, 1997) and appears to be the main storage site for spatial, verbal (Brownsett & Wise, 2010; Demonet, et al., 1994; Paulesu, et al., 2003), personal and even sexual memories (e.g., Gloor, 1997; Halgren, 1992). The temporal lobes, in fact, contain the core structures of the limbic system, the amygdala and hippocampus, and of all brain regions, only stimulation or activation of the temporal lobe or the underlying limbic structures, gives rise to personalized, subjective, emotional, and sexual experiences (Gloor, 1997; Halgren, 1992; Penfield & Perot, 1963). Stimulation of the temporal lobe can give rise to profound auditory or visual hallucinations, including the sensation or having left the body, as well as spiritual and religious feelings such as having the "truth" revealed and of receiving knowledge regarding the meaning of life and death.

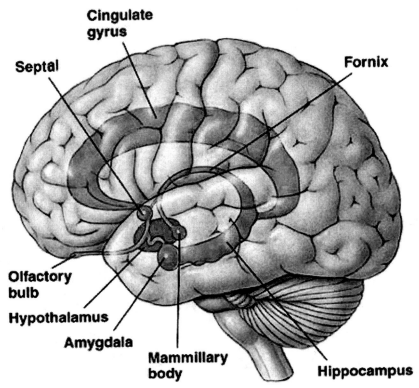

In addition to memory and personal and spiritual identity, the temporal lobe are responsive to complex auditory and visual stimuli (Binder et al., 1994; Gross & Graziano 1995; Nakamura et al. 1994; Nelken et al., 1999; Nishimura et al., 1999; Price, 1997; Rolls 1992; Tovee et al. 1994), and subserve the ability to comprehend complex and emotional speech. Functional imaging studies have repeatedly demonstrated activity in the superior and

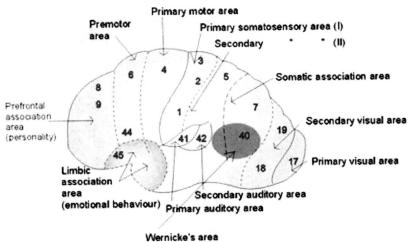

middle temporal lobe when engaged in language tasks (Bookheimer, et al., 1995; Bottini, et al., 1994; Fletcher et al., 1995; Howard et al., 1996; Shaywitz, et al., 1995; Warburton, et al., 1996). The temporal lobes, however, are functionally lateralized. Specifically, whereas the right temporal lobe subserves the ability to perceive and comprehend emotional, animal, environmental, and musical sounds (Joseph, 1988a; Parsons & Fox, 1997;

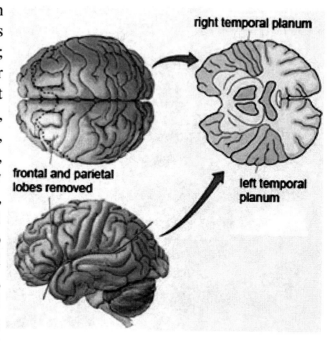

Ross, 1993) the left temporal lobe, including Wernicke's area is directly responsible for the capacity to comprehend complex human speech. The left superior temporal lobe, i.e. the planum temporal, which contains the auditory area, is in fact significantly larger than its counterpart on the right.

It has been determined, as demonstrated through lesion and physiological studies, that Wernicke's area (in conjunction with the inferior parietal lobule) acts to organize and separate incoming sounds into a temporal and interrelated series so as to extract linguistic meaning via the perception of the resulting sequences. When damaged a spoken sentence such as the "big black dog" might be perceived as "the klabgigdod." Patients develop what has been referred to as Wernicke's receptive aphasia. Presumably this disorder is due in part to an impaired capacity to discern the individual units of speech and their temporal order. That is, sounds must be separated into discrete interrelated linear units or they will be perceived as a meaningless blur.

Patients with damage to Wernicke's area are nevertheless, still capable of talking due to the preservation of Broca's area which continues to receive linguistic-related ideational material via the arcuate fasciculus pathway which originate in the temporal lobe and passes through the inferior parietal lobule--a structure which acts as a phonological storehouse that becomes activated during short-term verbal memory and word retrieval (Demonet, et al., 1994; Paulesu, et al., 1993; Price, 1997). However, because Wernicke's area is injured, speech output also becomes abnormal, a condition referred to as fluent aphasia. Broca's area keeps talking, but what is says is nonsense.

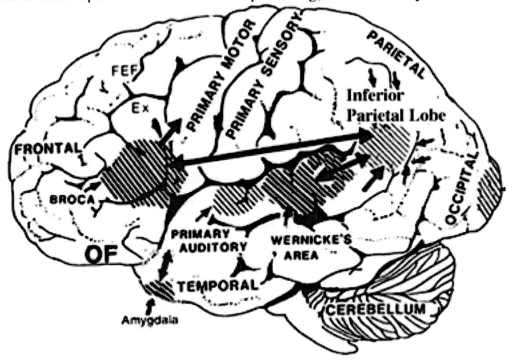

The speech output of a patient with "fluent aphasia" in many respects resembles the acutely psychotic speech of patients suffering from certain subtypes of schizophrenia. In fact, certain subtypes of schizophrenia have been repeatedly associated with abnormalities and irritative lesion involving the left temporal lobe (DeLisi et al. 1991; Dauphinais et al. 1990; Flor-Henry 1983; Perez et al. 1985; Rossi et al. 1990, 1991; Sherwin 1981). Conversely, given the role of the left temporal lobe in language, patients diagnosed with schizophrenia and who show signs of left temporal lobe dysfunction, also tend to demonstrate aphasic abnormalities in their thought and speech (Chaika, 1982; Flor-Henry, 1983; Hoffman, 1986; Hoffman, Stopek & Andreasen, 1986; Rutter, 1979).

The right temporal lobe also participates in language--as demonstrated through functional imaging (Bottini et al., 1994; Price et al., 1996; Shaywitz, et al., 1995)--and is especially responsive to sounds conveying emotion, melody, prosody, including those made by animals and those arising from the natural environment such as wind and rain (e.g. Fujii et al., 1990; Joseph, 1982, 1988a; Ross, 1993; Schnider et al. 1994; Zatorre & Halpen, 1993). It is the right temporal lobe which enables an individual to determine if someone is speaking sincerely, or with anger, happiness, and so on, whereas the left temporal lobe listens to the words being said.

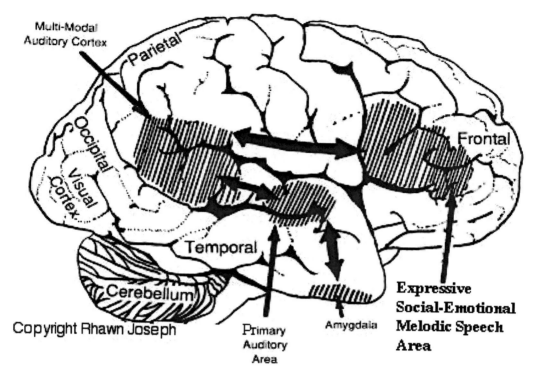

Multi-Modal
Auditory Cortex

Parietal

Occipital

Visual
Cortex

Frontal

Temporal

Cerebellum

Copyright Rhawn Joseph

Primary
Auditory
Area

Amygdala

**Expressive
Social-Emotional
Melodic Speech
Area**

Hence, if the right temporal lobe is severely injured, patients may suffer from a non-verbal social-emotional auditory agnosia (Joseph, 1982, 1988a, Ross, 1993; Schnider et al. 1994), also referred to as phonagnosia (van Lancker, et al., 1988), in which case they can no longer perceive social-emotional vocal nuances and may misperceive what others are saying or implying, and may be no longer capable of hearing "sincerity" or mirth or even "love" in which case they may become paranoid. Hence, although the ability to comprehend the non-emotional, denotative aspects of language is preserved (Fujii et al., 1990), patients with right superior temporal injuries may lose the capability to correctly discern environmental sounds (e.g. birds singing, doors closing, keys jangling), emotional-prosodic speech, as well as music (Nielsen, 1946; Ross, 1993; Samson & Zattore, 1988; Schnider et al. 1994; Spreen et al. 1965; Zatorre & Hapern, 1993).

The loss of the ability to appreciate music is due to the right temporal lobe being dominant for the perception of melodic and musical stimuli. For example, while listening to or performing scales, activity increases in the left temporal lobe, whereas when listening to Bach (the third movement of Bach's Italian concerto), the right temporal lobe becomes highly active (Parsons & Fox, 1997). Likewise, Evers and colleagues (1999) in evaluating cerebral blood velocity, found a right hemisphere increase in blood flow when listening to harmony (but not rhythm), among non-musicians in general, and especially among females. In fact, right temporal lobe activity increased when pianists were playing from memory Parsons & Fox, 1997). Conversely, right temporal injuries can disrupt the ability to remember musical tunes or to create musical imagery (Zatorre & Halpen, 1993).

Unfortunately, individuals with right temporal injuries not only lose the ability to appreciate music, but may misperceive and fail to comprehend a variety of paralinguistic social-emotional messages. This includes difficulty correctly identifying the voices of loved ones or friends, or discerning what others may be implying, or in appreciating emotional and contextual cues, including variables such as sincerity or mirthful intonation. In consequence, a patient may complain that his wife no longer loves him, and that he knows this from the sound of her voice. Or, because of difficulty discerning nuances such as humor and friendliness the patient may even become paranoid or delusional as they realize that friends and loved ones sound different, and may entertain delusions that they've been replaced by imposters.

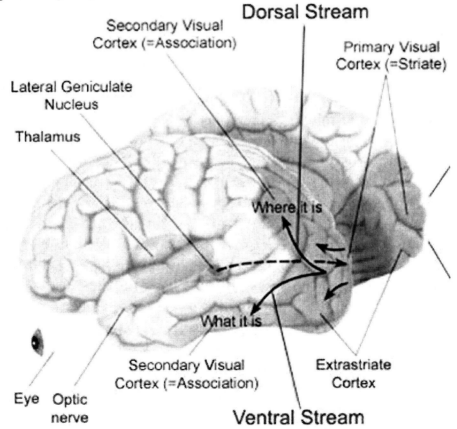

The temporal lobes also process complex visual stimuli, and can recognize faces, friends, and loved ones, as well as one's own face in the mirror. If injured, the ability to recognize faces or complex objects is compromised. Conversely, if abnormally activated, the result may be complex hallucinations.

The temporal lobes, therefore, are crucially important in memory, language, social emotional relations, maintaining personal identity and the ability to comprehend speech. If injured, patients may suffer memory loss, and/or language-related ideational activity may become abnormal, thus producing a formal thought disorder or delusions or paranoia, and personality functioning may become fractured and thus schizophrenic.

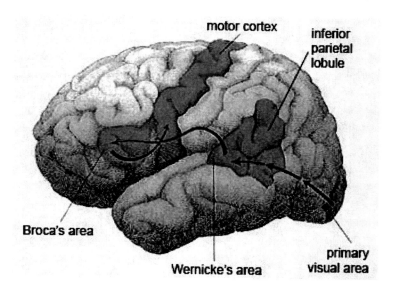

The Parietal Lobes

The parietal lobes maintain the body image and also consists of cells which are responsive to a variety of divergent stimuli, including movement, hand position, objects within grasping distance, audition, eye movement, pain, heat, cold, as well as complex and motivationally significant visual stimuli (Billington et al., 2010; Bisley & Goldberg 2010; Chong et al. 2010; Daprati et al., 2010; Gottlieb & Snynder 2010; Nelson et al. 2010; Seubert et al., 2008). The parietal lobes receives distinct sensory impressions from the entire body and can feel "pain" or a bug crawling on one's arm, leg, or face (Cohen & McCabe 2010; Daprati et al., 2010; Tsakiris 2010).The parietal lobes assimilate this information in order to coordinate reaching and the movements of the body in space, particularly the hand (Chong et al. 2010; Gentile et al., 2010). This is accomplished by transmitting this information to the motor areas of the frontal lobe, such that together, the frontal and parietal areas acts together to guide fine motor functioning.

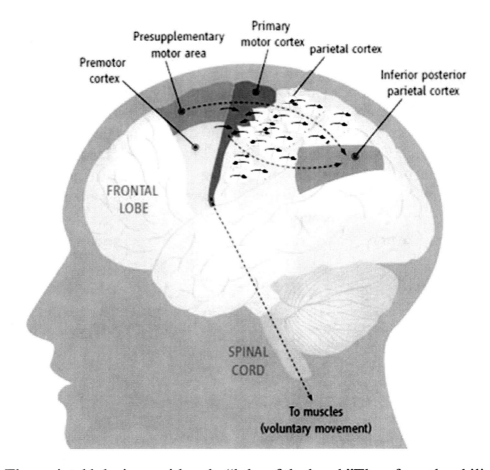

The parietal lobe is considered a "lobe of the hand." Therefore, the ability to reach for and manipulate a tool, open and remove the cap from a bottle and pour the contents into a glass, are made possible by the parietal lobe (Goldenberg et al. 2009; Ramayya et al. 2010) in concert with the frontal lobe which is dependent on the parietal somatosensory areas for tactile, skeletal, and visual feedback. Because of its importance in controlling hand movement, one consequence of parietal lobe injury may be apraxia--an inability to perform coordinated step-wise and sequential movements of the hands.

Patients with apraxia demonstrate gross inaccuracies as well as clumsiness when making reaching movements or when attempting to pick up small objects in visual space. They may also be impaired in their ability to acquire or perform tasks involving sequential changes in the hand or upper musculature including well learned, skilled, and even stereotyped motor tasks such as lighting a cigarette, using a key, or making a cup of coffee. That is, they may perform the various steps in the wrong order; e.g., pretending to stir the coffee, then pretending to pour it.

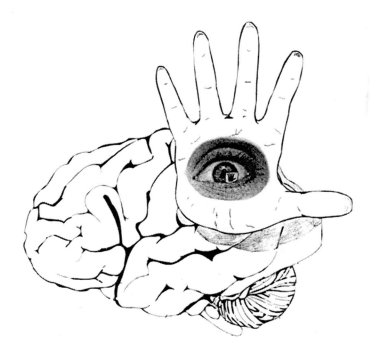

In addition, patients may no longer be able to perform constructional tasks, such as drawing or copying or performing puzzles or block designs, a condition referred to as constructional apraxia. They may leave out parts, grossly distort the figure, or even fail to draw half the object--particularly with right parietal injuries.

Apraxic disorders are most common with injuries involving the inferior parietal lobule (IPL), a structure that engages in the motor programming of hand movements, such as those involved in drawing, constructing, building, and which require sequential and orderly movements (De Renzi and Lucchetti, 1988; Heilman et al., 1982; Kimura, 1993; Strub and Geschwind, 1983). As noted above, the IPL (the angular and supramarginal gyrus) is also coextensive with Wernickes area and acts to assimilate auditory, visual, and tactile impressions, and provide names for these associated assimilation (which also makes reading and writing possible). Once this is accomplished, the IPL then injects this material, temporal sequential fashion, into the stream of language and thought all of which is transmitted to Broca's area and which is then expressed as grammatical speech (Joseph, 1982, 1986a, 1999e,f; Joseph et al., 1984).

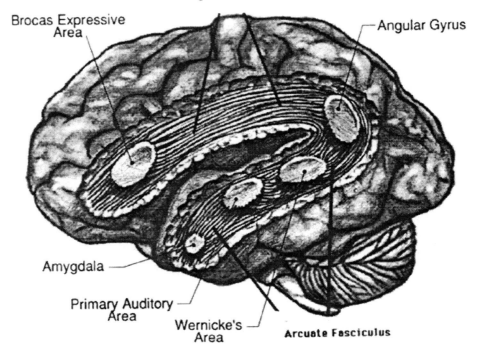

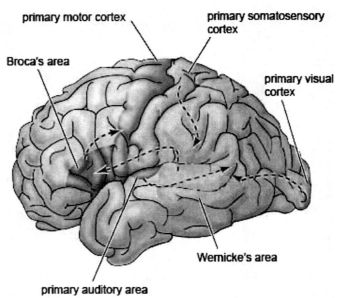

This is not merely a hypothesis based on lesion studies as these conclusions are supported by functional imaging, others have come to the same conclusion; i.e. the inferior parietal lobule acts as a phonological storehouse that becomes activated during short-term memory and word retrieval (Demonet, et al., 1994; Paulesu, et al., 1993; Price, 1997). For example, viewing words activates the left supramarginal gyrus (Bookheimer, et al., 1995; Vandenberghe, et al., 1996; , Menard, et al., 1996; Price, 1997) which also becomes active when performing syllable judgements (Price, 1997), and when reading (Bookheimer, et al., 1995; Menard, et al., 1996; Price, et al., 1996). Injuries to the IPL, therefore, can result in word finding difficulty (anomia) as well as a loss of the ability to read or write.

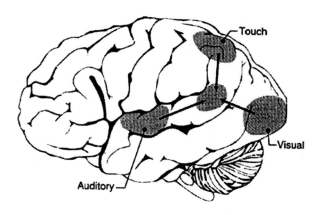

The parietal lobes also subserve and maintain a unique aspect of one's personal identity, the body image (Joseph, 1986a, 1988a). The parietal lobe actually maintains multiple maps of the body (Kaas, 1993; Lin et al., 1994), and in this regard, it is responsible for the ability to remember and recognize the body as an extension of one's personal self--particularly the right parietal lobe which appears to maintain multiple maps of both halves of the body and a bilateral image of body-image visual space (Joseph, 1986a, 1988a). Presumably, it is because the body image is maintained in the parietal lobe, that individuals who have suffered amputations continue to experience a phantom limb. Although the body part has been eliminated, the neural representation for the body may remain intact.

Because the parietal lobes maintain the body image, and as the entire body is multiply represented along the surface of the parietal lobes, massive injuries may result in a destruction of the body image. Memories of the body may be erased.

As noted, the right parietal lobe maintains a bilateral body image, the left parietal lobe appears to maintain a memory of only half the body (Joseph, 1986a, 1988a). If the left parietal lobe is injured, the right parietal area may continue to monitor both halves of the body and both halves of visual space and the body image will remain intact. If the right parietal area is severely injured, the left half of the body image, and in fact all memories of the left half of the body and the left half of space, may be abolished; a condition referred to as unilateral neglect. Patients may dress or groom only the right half of their body, eat only off the right half of their plates, and fail to read the left half of sentences and words, and so on. The left parietal lobe, having a memory of only the right half of the body, is unable to recognize the left half of the body, and ignores it and denies its existence.

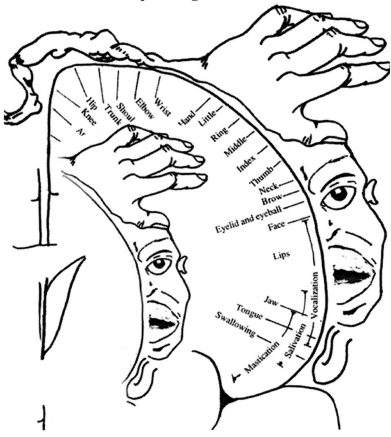

Following a massive right parietal injury, and when confronted with their unused or immobile limbs patients may (at least initially) deny that it belongs to them and instead claim it must belong to the doctor or a patient in the next bed. For example, Gerstmann (1942) describes a patient with left-sided hemiplegia who "did not realized and on being questioned denied, that she was paralyzed on the left side of the body, did not recognize her left limbs as her own, ignored them as if they had not existed, and entertained confabulatory and delusional ideas in regard to her left extremities. She said another person was in bed with her, a little Negro girl, whose arm had slipped into the patient's sleeve" (p. 894). Another declared, (speaking of her left limbs), "That's an old man. He stays in bed all the time."

With right parietal injuries coupled with unilateral neglect, patients may develop a dislike for their left limbs, try to throw them away, become agitated when they are referred to, entertain persecutory delusions regarding them, and even complain of strange people sleeping in their beds due to their experience of bumping into their left limbs during the night (Bisiach & Berti, 1987; Critchley, 1953; Gerstmann, 1942). One patient complained that the person tried to push her out of the bed and then insisted that if it happened again she would sue the hospital. A female patient expressed not only anger but concern least her husband should find out; she was convinced it was a man in her bed.

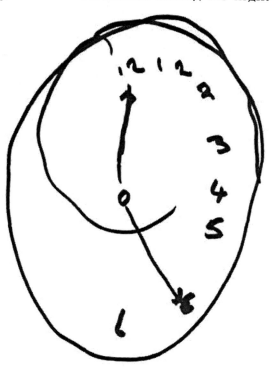

However, not just the left half of the body, but the left half of visual space may also be ignored and neglected.This can be demonstrated by asking the patient to draw the face of a clock and put all the numbers in.

The Occipital Lobes And Vision

The occipital lobe visual cortex performs complex visual analysis (Montemurro et al 2008; Priebe & Ferster 2008; Schwarzlose et al 2008). The occipital lobes are the smallest lobes of the brain, but like other tissues of the mind, they process information from a number of modalities and contain neurons which respond to vestibular, acoustic, visual, visceral, and somesthetic input (Beckers & Zeki 1994; Ferster, et al., 1996; Pigarev 1994; Priebe & Ferster 2008; Schwarzlose et al 2008). Predominantly, however, the neocortex of the occipital lobes are the main receiving stations for visual stimuli transmitted from the retina to the thalamus (Barbur et al., 1993; Ferster et al., 1996). Simple and complex visual and central/foveal analysis is one of the main functions associated with the occipital lobe (Montemurro et al 2008; Priebe & Ferster 2008; Schwarzlose et al 2008).

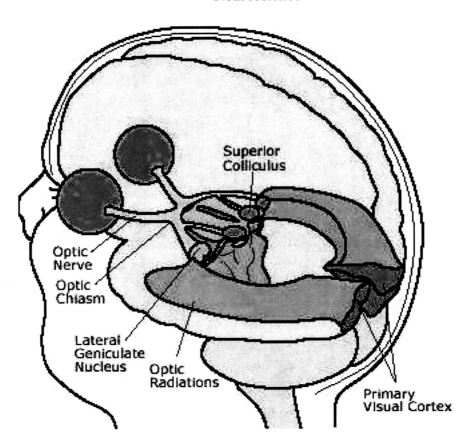

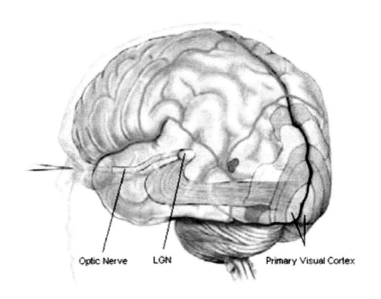

Visual Cortices

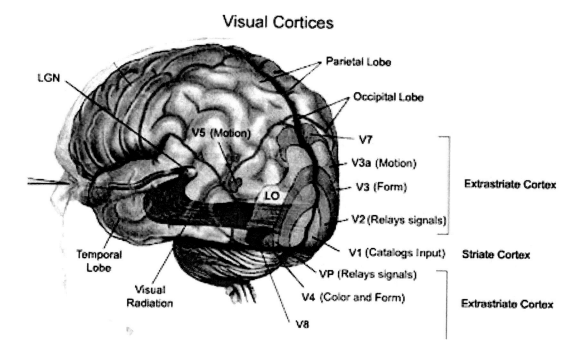

Specifically visual information is shunted from the lateral geniculate nucleus (LGN) of the thalamus to the primary visual receiving areas, striate cortex, area 17. Area 17 is referred to as "striate cortex" due to the striped appearance of layer IV, which is also directly innervated by the LGN. Layer IV is divided into three sublayers, with the middle layer containing a rather thick band of cortex, the band of Baillarger/Gennari, which is visible to the naked eye.

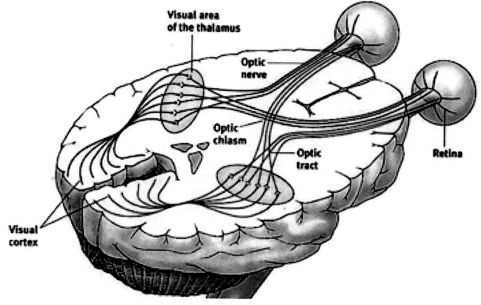

Throughout the striate cortex neurons with similar receptive properties are stacked in columns, with all the neurons within one column responding, for example, to a certain visual orientation, and the cells in the next column to an orientation of a slightly different angle. Columns exist for color, location, movement, etc, with some columns responding to input from one eye, i.e. ocular (eye) dominance columns (Hubel & Wiesel, 1968, 1974). In general, a strict topographical relationship is maintained throughout the visual projection system and the visual cortex. Within the visual cortex immediately adjacent groups of neurons respond to visual information from neighboring regions within the retina (Kaas & Krubitzer 1991).

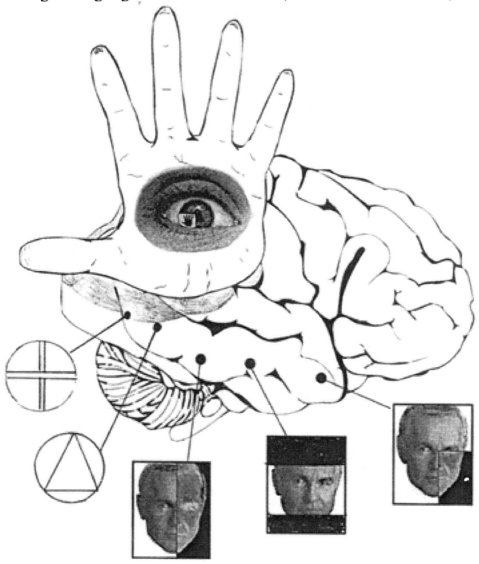

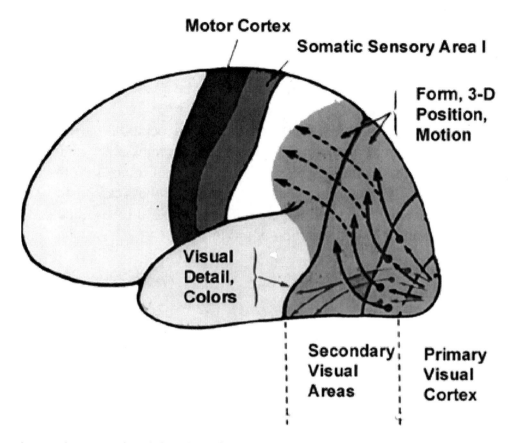

Information received in the visual areas of the occipital lobe are then projected dorsally and ventrally to a variety of visual association areas, including the parietal and temporal lobes (Kaas & Krubitzer 1991; Nakamura et al. 1994; Sereno et al. 1995; Tovee et al. 1994). The dorsal stream of visual information flows to and is assimilated by the parietal lobes and is incorporated for the purposes of coordinating body and arm and leg movements in visual space. The parietal (dorsal) visual stream, therefore is most sensitive to objects in the periphery and lower visual fields (Motter & Mountcastle, 1981; Previc 1990); i.e. where the hands, feet, and ground are more likely to be viewed, and which thus enables the parietal lobe to guide and observe the hands in motion.

The ventral visual stream flows from the occipital lobe to the inferior temporal lobe (ITL) where it is assimilated and also serves to activate feature detecting neurons which are sensitive to faces, objects, and other complex geometric stimuli. ITL neurons are sensitive to color, contrast, size, shape, orientation and are involved in the perception of three dimensional objects including specific shapes and forms including the human face (Eskander, et al. 1992; Gross & Graziano 1995; Gross, et al. 1972; Nakamura et al.

131

1994; Rolls, 1992; Sergent, et al.1990). In consequence, if injured, patients may suffer from an inability to recognize the faces of friends, loved ones, or pets (Braun et al. 1994; DeRenzi, 1986; Hanley et al. 1990; Hecaen & Angelergues, 1962; Landis et al., 1986; Levine, 1978); a condition referred to as prosopagnosia. Some patients may in fact be unable to recognize their own face in the mirror.

The ventral occipital/temporal and the dorsal occipital/parietal visual areas also interact. For example, the ventral stream tends to focus on objects, faces, and so on, whereas the dorsal stream focuses on the hands in visual space. By interacting, the hands (parietal lobes) can be directed at specific objects identified and targeted by the temporal lobes, thus making coordinated hand-eye coordination possible.

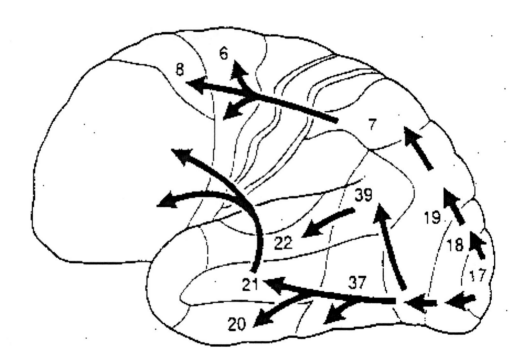

The Primary, Secondary And Association Areas

The mosaic which can loosely be defined as consciousness, consists of multiple and parallel streams of information which are processed hierarchically, horizontally, vertically, and in modular fashion, with the brainstem, cerebellum, diencephalon, striatum, limbic system, and neocortex providing their own unique contributions. Conscious processing within the neocortex also occurs hierarchically, vertically, horizontally, as well as in parallel, even within the separate lobes of the brain.

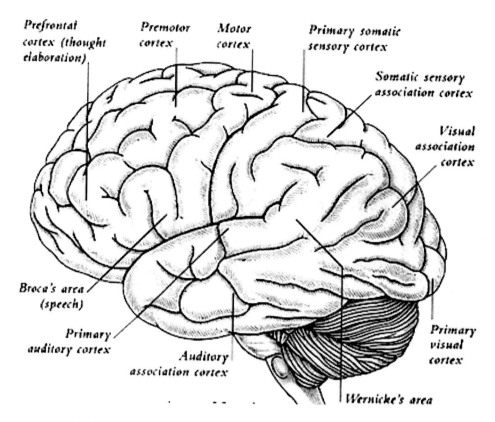

Prefrontal cortex (thought elaboration)

Premotor cortex

Motor cortex

Primary somatic sensory cortex

Somatic sensory association cortex

Visual association cortex

Broca's area (speech)

Primary auditory cortex

Auditory association cortex

Primary visual cortex

Wernicke's area

For example, in addition to their layered and columnar organization (Mountcastle, 1997), each of the four lobes of the brain consists of multiple cellular cytoarchitectonic subdivisions which perform unique and/or overlapping functions. For example, each lobe can be subdivided into primary, secondary, association, and assimilation areas. In general, information processing in the sensory receiving areas of the parietal, temporal, and occipital lobes begins in the primary receiving area and then flows to the secondary receiving area and then to the association and multimodal assimilation areas (Jones & Powell, 1970; Pandya & Yeterian, 1985). However, since the various subdivisions of the thalamus also project to the association areas, these tissues of the mind may receive specific types of visual, or auditory, or somesthetic input in advance of the primary areas (e.g. Zeki, 1997).

In general, however, the flow is from primary to secondary to association area. For example, somatosensory information is first transmitted from the thalamus to the primary receiving areas of the parietal lobe (Brodmann's cytoarchitectonic areas 3,2,1). Here the individual sensory elements are analyzed and localized, e.g., cold, wet, hard, small, cubicle, palm. These sensory impressions are then shunted to the secondary areas (Brodmann's area 5) where these impressions may be combined, e.g. an ice cube held in the hand. These association areas are then transferred to the association-assimilation areas (Brodmann's areas 7, 39, 40), i.e. the inferior parietal lobule where they may then be visualized and even named.

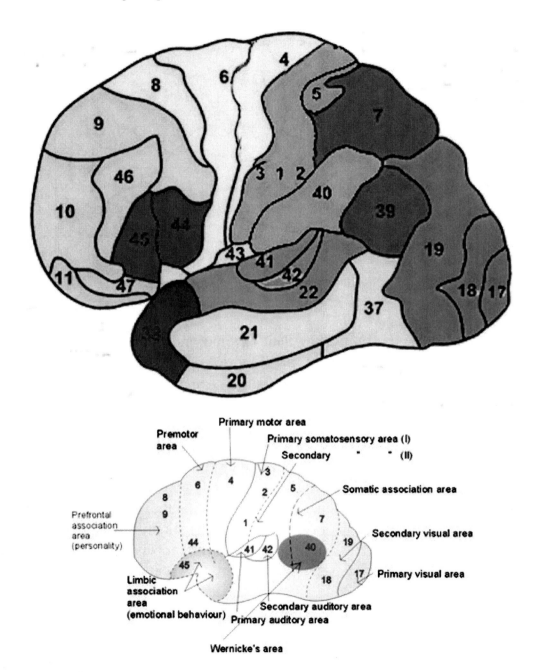

As per the motor areas, the opposite sequence of events takes places, with information flow beginning in the supplementary motor areas (SMA)

prior to movement, and prior to activation of the secondary-association areas-- cytoarchitectonic areas 8,6 (Alexander & Crutcher, 1990; Crutcher & Alexander, 1990). The SMA may well initiate movements as this part of the brain has also been associated with what has been described as the "will."

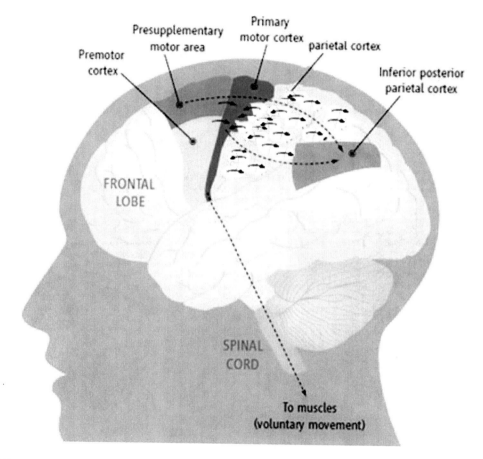

The secondary motor areas also become activated prior to movement and precedes cellular activation of the primary region (Alexander & Crutcher, 1990; Crutcher & Alexander, 1990; Weinrich et al. 1984). The premotor area appears to program various gross and fine motor activities, and becomes highly active during the learning of new motor programs (Porter 1990; Roland et al. 1981).

Depending on the task, the primary motor areas may show low level activity prior to movement, but then becomes highly active during movement (Passingham, 1999)--a reflection of the fact that it supplies at least a third of the axonal trunk which makes up the corticospinal tract and whose cells maintain multiple musculature connections (via the spinal cord), as well as a one to one correspondence with specific muscles. In fact, direct electrical stimulation of the frontal primary motor cortex can induce twitching of the lips, flexion or extension of a single finger joint, protrusion of the tongue or elevation of the palate (Penfield & Boldrey, 1937; Penfield & Jasper, 1954; Penfield & Rasmussen, 1950; Rothwell et al. 1987), even though patients never claim to have willed these movements. Again, however, like the sensory areas, different motor subareas may become activated simultaneously, including the primary motor area which may become aroused prior to movement, only to increase in activity during specific movements (Passingham, 1993).

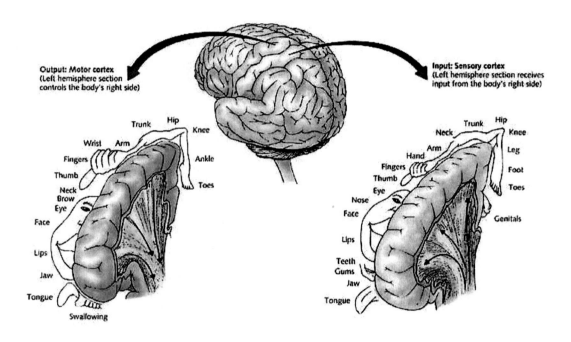

However, the motor areas also interact with the sensory areas in the parietal lobe. In this way, motor movement can be coordinated with sensory impressions arising from the skin, joints, muscles, and body.

Incoming and outgoing signals, therefore, are processed sequentially and in parallel. Beginning in the primary receiving areas these signals are relayed via a variety of separate branches to the secondary receiving areas and to association areas further downstream which are also simultaneously receiving thalamic input, processing this data, and then relaying this information back to the primary areas as well as laterally to other association areas (Kaas, 1993; Panya & Yeterian, 1985; Zeki, 1997) and back to the thalamus and to the frontal lobes. A similar stepwise and parallel process occurs; albeit in reverse, within the motor areas. That is, the primary motor areas in the frontal lobe may also become active prior to movements, but then increase their activity when the movement is actually performed (Passingham, 1997).

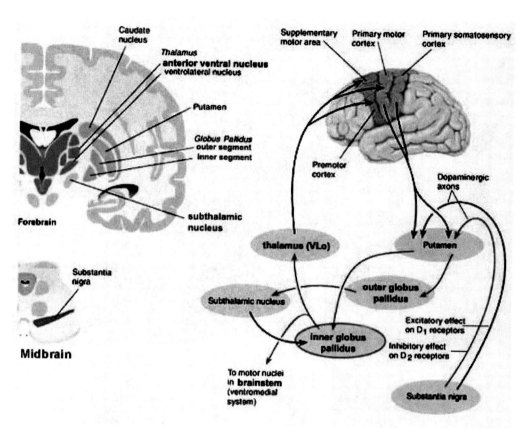

Hence, these divergent cytoarchitectonic areas do not initially act in isolation as multiple areas in diverse regions of the brain become activated simultaneously and interact in regard to perceptual and motor functioning. That is, functions are localized as well as distributed throughout the neuroaxis with different regions making unique and overlapping contributions to the mosaic of the mind.

The Split-Brain. Right And Left Hemisphere: Functional Laterality

Those aspects of the mind that we associate with consciousness, language, and rational thought, are clearly maintained and propagated by the neocortical mantle of the frontal, temporal, parietal, and occipital lobes and is divisible horizontally, vertically, hierarchically, cytoarchitecturally, and in accordance with the specialities of the different lobes of the brain including the limbic system and diencephalon. These different temporal and parallel streams of mental activity, including those generating by the structures of the limbic system, coalesce to create a conscious composite, a semi-integrated whole.

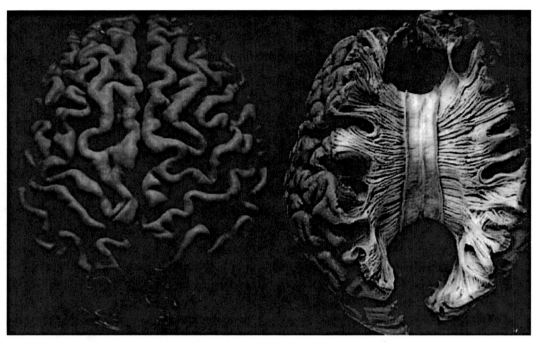

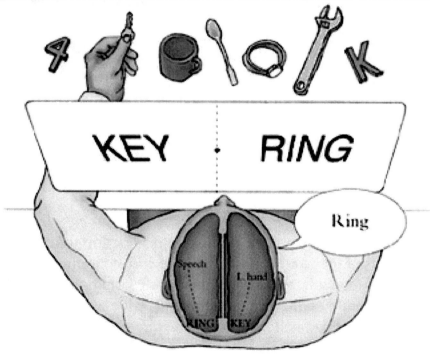

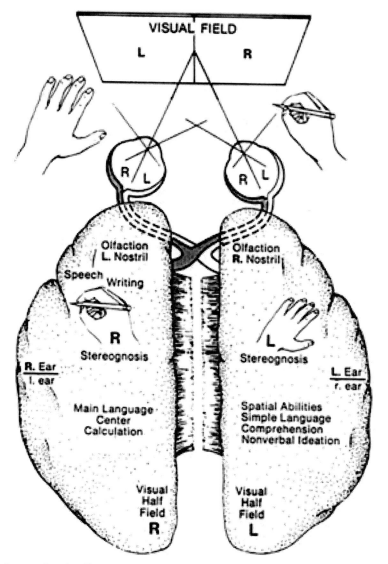

The cerebrum including its striatal, limbic, and diencephalic components, is divisible not only horizontally, vertically, hierarchically, and in regard to its different lobes, but laterally. That is, two semi-independent streams of conscious awareness coexist literally side by side. Each half of the cerebrum, which includes neocortical, striatal, limbic and diencephalic structures, has its own memories, goals, personal identity, and social emotional orientation and capabilities or lack-thereof, as has now been repeatedly demonstrated following surgical sectioning of the thick interhemispheric axonal pathway, the corpus callosum (Akelaitis, 1945; Bogen, 1969, 1979; Joseph, 1986b, 1988b; Levy, 1974, 1983; Sperry, 1966, 1982).

As described by Nobel Lauriate Roger Sperry (1966, p. 299), "Everything we have seen indicates that the surgery has left these people with two separate minds, that is, two separate spheres of consciousness. What is experienced in the right hemisphere seems to lie entirely outside the realm of awareness of the left hemisphere. This mental division has been demonstrated in regard to perception, cognition, volition, learning and memory."

In consequence, following the surgical sectioning of the corpus callosum, each half of the brain and mind may act independently. For example, Akelaitis (1945, p. 597) describes patients with complete corpus callosotomies who experienced extreme difficulties making the two halves of their bodies and the two halves of their brains cooperate, as each half of the brain apparently had its own desires, goals and intentions. A recently divorced male "split-brain" patient noted that on several occasions while walking about town he found himself forced to go some distance in another direction by the left half of his body. Specifically, whereas the speaking, language dominant left half of his brain just wanted to go for a walk, the right half of his brain was trying to walk toward the new home of his former wife.

A "split-brain" patient I examined, who had recently broken up with his girlfriend, indicated thumbs down and stated matter-of-factly that he had absolutely no desire to see her again. However, when his (non-verbal) right hemisphere was asked how it felt about the situation, and was told to give a thumbs up or down, it gave a thumbs up (with the left hand) when asked if he still wanted to see her.

The right hemisphere of another split-brain patient apparently felt considerable animosity for his wife, for it slapped her several times---- much to the embarrassment of his left (speaking) hemisphere. In another case, a patient's left hand attempted to choke the patient himself and had to be wrestled away. The right and left hemisphere of yet another split-brain patient enjoyed different television programs, different foods, and had different attitudes even about mundane activities such as going for a walk (Joseph, 1988b).

For example, while watching television one afternoon, this patient "2-C" reported that to the dismay of his left (speaking) hemisphere, he was dragged from the couch by his left leg, and that the left half of his body dragged him to the TV where his left hand then changed channels even though he (or rather his left hemisphere) was enjoying the program. On another occasion, it simply turned the TV off and tried to leave the room. Once, after he had retrieved something from the refrigerator with his right hand, his left hand took the food, put it back on the shelf and retrieved a completely different item, "Even though that's not what I wanted to eat!" his left hemisphere complained. On at least one occasion, his left leg refused to continue "going for a walk" and would only allow him to return home.

2-C was so annoyed with the independent actions of the left half of his body that he frequently expressed hate for it, even striking it angrily with the right hand. Nevertheless, the right hemisphere knew exactly what it was doing, as was demonstrated experimentally, and thus had its own goals, desires, intentions, and favorite foods and TV shows--even showing the good sense of turning off the TV and leaving the room (Joseph, 1988b).

Dissociation and Self-Consciousness

The multiplicity of mind is not limited to the neocortex but includes old cortical structures, such as the limbic system (Joseph 1992). Moreover, limbic nuclei such as the amygdala and hippocampus interact with neocortical tissues creating yet additional mental systems, such as those which rely on memory and which contribute to self-reflection, personal identity, and even self-consciousness (Joseph, 1992, 1998, 1999b, 2001).

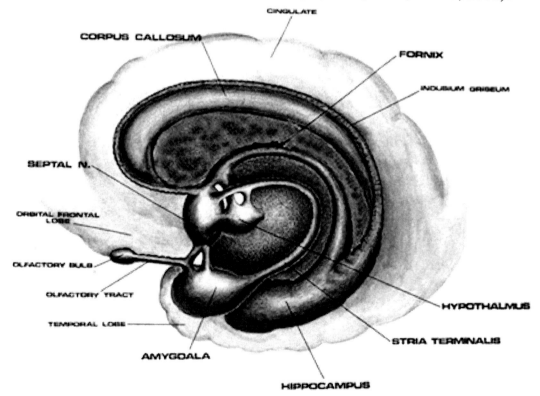

For example, both the amygdala and the hippocampus are implicated in the storage of long term memories, and both nuclei enable individuals to visualize and remember themselves engaged in various acts, as if viewing their behavior and actions from afar. Thus, you might see yourself and remember yourself engage in some activity, from a perspective outside yourself, as if you are an external witness; and this is a common feature of self-reflection and self-memory and is made possible by the hippocampus and overlying temporal lobe (Joseph, 1996, 2001).

The hippocampus in fact contains "place neurons" which cognitive map one's position and the location of various objects within the environment (Nadel, 1991; O'Keefe, 1976; Wilson & McNaughton, 1993). Further, if the subject moves about in that environment, entire populations of these place cells will fire. Moreover, some cells are responsive to the movements of other people in that environment and will fire as that person is observed to move around to different locations or corners of the room (Nadel, 1991; O'Keefe, 1976; Wilson and McNaughton, 1993).

Electrode stimulation, or other forms of heightened activity within the hippocampus and overlying temporal lobe can also cause a person to see themselves, in real time, as if their conscious mind is floating on the ceiling staring down at their body (Joseph, 1998, 1999b, 2001). During the course of electrode stimulation and seizure activity originating in the temporal lobe or hippocampus, patients may report that they have left their bodies and are hovering upon the ceiling staring down at themselves (Daly, 1958; Penfield, 1952; Penfield & Perot 1963; Williams, 1956). That is, their consciousness and sense of personal identity appears to split off from their body, such that they experience themselves as as a consciousness that is conscious of itself as a conscious that is detached from the body which is being observed.

One female patient claimed that she not only would float above her body, but would sometimes drift outside and even enter into the homes of her neighbors. Penfield and Perot (1963) describe several patients who during a temporal lobe seizure, or neurosurgical temporal lobe stimulation, claimed they split-off from their body and could see themselves down below. One woman stated: "it was though I were two persons, one watching, and the other having this happen to me." According to Penfield (1952), "it was as though the patient were attending a familiar play and was both the actor and audience."

Under conditions of extreme trauma, stress and fear, the amygdala, hippocampus and temporal lobe become exceedingly active (Joseph, 1998, 1999b). Under these conditions many will experience a "splitting of consciousness" and have the sensation they have left their body and are hovering beside or above themselves, or even that they floated away (Courtois, 2009; Grinker & Spiegel, 1945; Noyes & Kletti, 1977; van der Kolk 1987). That is, out-of-body dissociative experiences appear to be due to fear induced hippocampus (and amygdala) hyperactivation.

Likewise, during episodes of severe traumatic stress personal consciousness may be fragmented and patients may dissociate and experience themselves as splitting off and floating away from their body, passively observing all that is occurring (Courtois, 1995; Grinker & Spiegel, 1945; Joseph, 1999d; Noyes & Kletti, 1977; Southard, 1919; Summit, 1983; van der Kolk 1987).

Noyes and Kletti (1977) described several individuals who experienced terror, believed they were about to die, and then suffered an out-of body dissociative experience: "I had a clear image of myself... as though watching it on a television screen." "The next thing I knew I wasn't in the truck anymore; I was looking down from 50 to 100 feet in the air." "I had a sensation of floating. It was almost like stepping out of reality. I seemed to step out of this world."

One individual, after losing control of his Mustang convertible while during over 100 miles per hour on a rain soaked freeway, reported that "time seemed to slow down and then... part of my mind was a few feet outside the car zooming above it and then beside it and behind it and in front of it, looking at and analyzing the respective positions of my spinning Mustang and the cars surrounding me. Simultaneously I was inside trying to steer and control it in accordance with the multiple perspectives I was given by that part of my mind that was outside. It was like my mind split and one consciousness was inside the car, while the other was zooming all around outside and giving me visual feedback that enabled me to avoid hitting anyone or destroying my Mustang."

Numerous individuals from adults to children, from those born blind and deaf, have also reported experiencing a dissociative consciousness after profound injury causing near death (Eadie 1992; Rawling 1978; Ring 1980). Consider for example, the case of Army Specialist J. C. Bayne of the 196th Light Infantry Brigade. Bayne was "killed" in Chu Lai, Vietnam, in 1966, after being simultaneously machine gunned and struck by a mortar. According to Bayne, when he opened his eyes he was floating in the air, looking down on his burnt and bloody body: "I could see me... it was like looking at a manikin laying there... I was burnt up and there was blood all over the place... I could see the Vietcong. I could see the guy pull my boots off. I could see the rest of them picking up various things... I was like a spectator... It was about four or five in the afternoon when our own troops came. I could hear and see them approaching... I looked dead... they put me in a bag... transferred me to a truck and then to the morgue. And from that point, it was the embalming process. I was on that table and a guy was telling jokes about those USO girls... all I had on was bloody undershorts... he placed my leg out and made a slight incision and stopped... he checked my pulse and heartbeat again and I could see that too... It was about that point I just lost track of what was taking place.... [until much later] when the chaplain was in there saying everything was going to be all right.... I was no longer outside. I was part of it at this point" (reported in Wilson, 1987, pp 113-114; and Sabom, 1982, pp 81-82).

Therefore, be it secondary to the fear of dying, or depth electrode stimulation, these experiences all appear to be due to a mental system which enables a the conscious mind to detach completely from the body in order to make the body an object of consciousness (Joseph, 1998, 1999b, 2001).

The Neuroanatomy Of Mind

In summary, the human mind and brain are functionally lateralized, sexually differentiated, and hierarchically, vertically, and horizontally organized, and are significantly shaped and effected by experience. Considered broadly, it can be said that the most unconscious and reflexive aspects of the "unconscious" mind are associated with the brainstem.

The diencephalon, which is immediately anterior and adjacent to the brainstem, is associated with a vague sensory affective unconscious, such that pre-conscious cognitive-sensory processing (thalamus) and reflexive emotional processing (hypothalamus) takes place in this region-- information which may be relayed to the neocortex as well as to the limbic system.

The limbic system, which is anterior to the diencephalon, and is both dorsally and ventrally situated, is capable of exceedingly complex and sophisticated mental activity and can process, analyze, and learn and remember complex cognitive, linguistic, visual-spatial, and affective material, as well as generate complex emotions ranging from love to hate. The limbic system can also vocalize and it can think, and it can transfer this information to the overlying neocortex which it may impel to act on its desires and its fears.

The neocortex is associated with the more rational and logical aspects of the mind. It is this neocortical shroud which envelops and coats the cerebrum with six to seven layers of gray matter, which cogitates, speaks in words and sentences, and can reason, rhyme, plan for the future, as well as ponder and analyze its own brain and mind.

The mind, however, is also a continuum, and the human brain is a composite of interacting structures which are intimately interconnected and which perform a variety of unique and overlapping functions often in parallel. And yet, although these structures interact and often engage in parallel processing, they are also functionally specialized, with some areas, such as the brainstem and limbic system, often acting completely independently of the neocortex or what is classically referred to as the conscious mind.

Overview: Consciousness, Awareness, And The Neuroscience Of Mind

The brain and the mind are synonymous, for if the brain is damaged, so too is the mind. The brain and associated mental activity are functionally lateralized and hierarchically organized, such that specific cognitive and emotional functions may be localized to specific regions of the brain. There are phylogenetically older, sensory-motor non-emotional mental systems associated with the brainstem, and a highly complex emotional and memory system associated with the limbic system.

Whereas the brainstem is incapable of conscious or cognitive activity, and instead mediates reflexive motor acts, including breathing, heart rate and arousal, it is the limbic system which mediates the ability to feel love or sorrow or to determine if something is good to eat. Indeed, it is the limbic system which enables humans to form memories of the long ago, as well as to recall these memories in order to dream of the future, as specific limbic nuclei become active not only when learning and remembering, but when dreaming.

In fact, the limbic system provides humans (and perhaps non-humans) with the capacity to experience the most profound of emotions, from love to spiritual ecstasy and religious awe, serving, at its most profound, as perhaps even a transmitter to god.

By contrast, the more recently evolved neocortex of the right hemisphere is the domain of a highly evolved social-emotional, visual-pictorial, spatial, body centered awareness and employs emotional and melodic sounds for expression. The right half of the brain is responsible for discerning distance, depth and movement, and recognizing environmental and animal sounds such as a chirping bird, a buzzing bee, a babbling brook, or a thunderstorm, as well as the capacity to sing, dance, chase or throw something with accuracy, and run without falling or bumping into things. Whereas the left hemisphere is concerned with logic and grammatical rules of organization and expression, including the analysis of details, sequential units, and parts, it is the right hemisphere which is able to perceive events or stimuli as a whole and which can see the forest as well as the trees. Whereas right hemispheric activity is associated with the production of the visual, emotional, hallucinatory, hypnogogic aspects of dreaming, the left hemisphere tends to immediately forget the dream upon waking (Joseph, 1988a).

Hence, in the right hemisphere we deal with a non-verbal form of awareness that accompanies in parallel the temporal-sequential, language dependent stream of consciousness associated with the functional integrity of the left cerebrum. In fact, and as has been repeatedly demonstrated, the right and left cerebral hemisphere are each capable of self-awareness, can plan for the future, have goals and aspirations, likes and dislikes, social and political awareness, and can independently and purposefully initiate behavior, guide responses choices and emotional reactions, as well as recall and act upon certain desires, impulses situations or environmental events --sometimes without the aid, knowledge or active (reflective) participation of the other half of the brain.

Thus, in summary, the right and left hemisphere subserve almost wholly different mental systems--a function of evolutionary metamorphosis and the evolution of language and tool making capabilities-- whereas the limbic system is synonymous with the most archaic regions of the emotional, sexual, unconscious mind. However, because the brain and mind are hierarchically organized and lateralized, and as the limbic system retains the capacity to completely overthrow and hijack the neocortically equipped "rational" and emotionally intelligent mind, conscious-awareness, as well as the brain, are subject to fracture, with each isolated segment acting on its own wishes and desires, independent of and sometimes conflicting with those mental systems that may remain intact such as those associated with the language-dependent conscious mind.

REFERENCES

For a complete bibliography , See Neuroscience, Neuropsychology, Neuropsychiatry, Brain and Mind, 4th edition, by R. Joseph

Abecasis, D. et al., (2009). Brain Lateralization of Metrical Accenting in Musicians, Annals of the New York Academy of Sciences. 1169, 74-78.

Aeschbach, D. (2011). REM-Sleep Regulation. In Mallick, B. N. et al. (eds) Rapid Eye Movement Sleep: Regulation and Function, Cambridge University Press.

Akiyam, T., & Tsuchiya, M. (2009). Study on pathological mechanisms of temporal lobe epilepsy and psychosis through kindling effect, Asian Journal of Psychiatry, 2, 37-39.

Alexander, M. P., & Hillis, A. E. (2008). Aphasia Handbook of Clinical Neurology, 88, 287-309.

Anders, S. et al. (2009). When seeing outweighs feeling: a role for prefrontal cortex in passive control of negative affect in blindsight.

Alpers, G. W. (2008). Eye-catching: Right hemisphere attentional bias for emotional pictures. Laterality: Asymmetries of Body, Brain and Cognition, 13, # 2. 10.1080/13576500701779247

Ardila, A. (2010). Proposed reinterpretation and reclassification of aphasic syndromes. Aphasiology, 24, 363-394.

Arnsten, A. F. T. (2009). Stress signalling pathways that impair prefrontal cortex structure and function, Nature Rev. Neuroscience, 10, 410-422.

Avidan, G. & Behrmann, M. (2009). Functional MRI Reveals Compromised Neural Integrity of the Face Processing Network in Congenital Prosopagnosia. Current Biology, 19, 1146-1150.

Barbeau. E. J. et al., (2008). Spatio temporal dynamics of face recognition. Cerebral Cortex, 18, 997-1009.

Badre D, Wagner AD. (2005). Frontal lobe mechanisms that resolve proactive interference. Cereb. Cortex 15:2003–12.

Baier B, Karnath HO (2008) Tight link between our sense of limb ownership and self-awareness of actions. Stroke 39:486–488.

Baier, B. et al. (2010).Keeping memory clear and stable—the contribution of human basal ganglia and prefrontal cortex to working memory J. Neuroscience, 30, 9788-9782

Ballmaier, M. et al. (2008). Regional patterns and clinical correlates of basal ganglia morphology in non-medicated schizophrenia, Schizophrenia Research, 106, 140-147.

Barlow, J. S. (2005). The Cerebellum and Adaptive Control, Cambridge University Press.

Basso, D. et al., (2010). Prospective memory and working memory: Asymmetrical effects during frontal lobe TMS stimulation, Neuropsychologia, 48, 3282-3290.

Beaucousin, V., et al. (2007). FMRI study of emotional speech comprehension. Cereb. Cortex 17, 339–352.

Becker, J. B. et al. (2007). Sex Differences in the Brain: From Genes to Behavior . Oxford University Press.

Bédard, P. & Sanes, J. N. (2011). Basal ganglia-dependent processes in recalling learned visual-motor adaptations, Experimental Brain Research, 209. 385-393.

Berryhill, M., & Olson, I. R. (2008). The right parietal lobe is critical for visual working memory. Neuropsychologia, 46, 1767-1774

Berthier, M. L., & Pulvermuller, F. (2011). Neuroscience insights improve neurorehabilitation of poststroke aphasia, Nature Reviews Neurology 7, 86-97.

Bever, T. G., & Chiarello, R. J. (2009). Cerebral dominance in musicians and nonmusicians, Journal of Neuropsychiatry, 21, 94-97.

Bhatt, S. et al (2005). Cytokine modulation of defensive rage behavior in the cat: role of GABAA and interleukin-2 receptors in the medial hypothalamus, Neuroscience, 133, 17-28.

Bignal, J-P, et al., (2007). The dreamy state: hallucinations of autobiographic memory evoked by temporal lobe stimulations and seizures. Brain 130, 88-99.

Billington, J., et al., (2010). An fMRI study of parietal cortex involvement in the visual guidance of locomotion. Journal of Experimental Psychology: Human Perception and Performance, Vol 36(6), Dec 2010, 1495-1507.

Binder, J. R. et al., (2010) Mapping anterior temporal lobe language areas with fMRI: A multi-center normative study. Neuroimage, 54, 1465-1475.

Bisley, J. W. & Goldberg, M. E. (2010). Attention, Intention, and Priority in the Parietal Lobe. Annual Review of Neuroscience, 33, 1-21.

Blanke O, Metzinger T (2009) Full-body illusions and minimal phenomenal selfhood. Trends Cogn Sci 13:7–13

Blanke O, Mohr C (2005) Out-of-body experience, heautoscopy, and autoscopic hallucination of neurological origin implications for neurocognitive mechanisms of corporeal awareness and self- consciousness. Brain Res Brain Res Rev 50:184–199

Bookheimer, S. (2002). Functional MRI of language: new approaches to under- standing the cortical organization of semantic processing. Annu. Rev. Neurosci. 25, 151–188.

Bright, P., Moss, H., Tyler, L.K., (2004). Unitary vs multiple semantics: PET studies of word and picture processing. Brain Lang. 89, 417–432.

Brown, G. M. et al. (2010). The role of melatonin in seasonal affective disorder. In Partonen. T., & Pandi-Perumal, S. R. (Eds). Seasonal affective disorder: practice and research, Oxford University Press.

Brownsett, S. L. E. & Wise, R. J. S. (2010). The Contribution of the Parietal Lobes to Speaking and Writing. Cerebral Cortex, 20, 517-523.

Buch, E. R. et al., (2010). A network centered on ventral premotor cortex exerts both facilitatory and inhibitory control over primary motor cortex during action reprogramming, J. Neuroscience, 30, 4882-4889.

Budson, A., & Kowall, N. W,. (2011). The Handbook of Alzheimer's Disease and Other Dementias, Wiley-Blackwell.

Burman, D. B., et al., (2008). Sex differences in neural processing of language among children, Neuropsychologia, 46,1349-1362.

Buxbaum, L.J., et al., (2007). Left inferior parietal representations for skilled hand-object interactions: evidence from stroke and corticobasal degeneration, Cortex, 43, 411-423.

Cabeza R, Dolcos F, Graham R, Nyberg L. (2002). Similarities and differences in the neural correlates of episodic memory retrieval and working memory. Neuroimage 16:317–30.

Cappelletti, M., et al., (2010). The Role of Right and Left Parietal Lobes in the Conceptual Processing of Numbers. Journal of Cognitive Neuroscience, 22, 331-346.

Cartwright, R. D. (2007). Night Life: Explorations in Dreams. BookSurge Publishing.

Caspers, S. et al., (2008). The human inferior parietal lobule in stereotaxic space, Brain, Structure, Function, 212, 481-495.

Cauquil-Michon, C. et al. (2011). Borderzone Strokes and Transcortical Aphasia, Current Neurology and Neuroscience Reports, 11, 570-577.

Chambers, C. D. et al. (2006). Executive "Brake Failure" following Deactivation of Human Frontal Lobe, Journal of Cognitive Neuroscience, 18, 444-455.

Chen, J. et al. (2010). Switching to Hypomania and Mania: Differential Neurochemical, Neuropsychological, and Pharmacologic Triggers and Their Mechanisms, Current Psychiatry Reports, 12, 512-521.

Chen, Y. H., et al., (2011). Increased risk of schizophrenia following traumatic brain injury: a 5-year follow-up study in Taiwan. Psychological Medicine. 22:1-7.

Chong, T. T. J. et al., (2010). fMRI Adaptation Reveals Mirror Neurons in Human Inferior Parietal Cortex. Current Biology, 18, 1576-1580.

Chouinard, P. A., Paus, T. (2006). The Primary Motor and Premotor Areas of the Human Cerebral Cortex, Neuroscientist April 2006 vol. 12 no. 2 143-152.

Chow, E. W. C. et al., (2011). Association of Schizophrenia in 22q11. 2 Deletion Syndrome and Gray Matter Volumetric Deficits in the Superior Temporal Gyrus. American Journal of Psychiatry, 168, 522-529.

Cleret de Langavant L, Trinkler I, Cesaro P, Bachoud-Lévi AC (2009) Heterotopagnosia: when I point at parts of your body. Neuropsychologia 47:1745–1755.

Cohen, H. and McCabe, C., 2010. A case of complex regional pain syndrome with extensive severe allodynia, referred sensations and clinical evidence of parietal lobe dysfunction: cortical reorganisation contributes to clinical presentation. Rheumatology, 49 (Supplement 1), I40-I40.

Collinson, S. L., et al. (2009). Dichotic listening impairments in early onset schizophrenia are associated with reduced left temporal lobe volume, Schizophrenia research, 112, 24-31.

Chun, W. & Johnson, G.V. (2007). The role of tau phosphorylation and cleavage in neuronal cell death. Front Biosci 12: 733–56.

Clark, R. E. & Squire, L. (2010). An animal model of recognition memory and medial temporal lobe amnesia: History and current issues, Neuropsychologia, 48, 2234-2244.

Cohen, R., et al. (2007). Notation-dependent and -independent representations of numbers in the parietal lobes, Neuron, 53, 307-314.

Correia, S., et al. (2010) Basal Ganglia MR Relaxometry in Obsessive-Compulsive Disorder: T2 Depends Upon Age of Symptom Onset, Brain imaging and Behavior, 4, 35-45.

Cowey, A. (2010) The blindsight saga. Exp. Brain Res. 200, 3-24.

Craig AD (2009) How do you feel-now? The anterior insula and human awareness. Nat Rev Neurosci 10:59–70

Cui, L. Q. et al. (2010). A comparative study of voxel-based morphometry in patients with paranoid schizophrenia and bipolar mania, Sichuan Da Xue Xue Bao Yi Xue Ban. 2010 Jan;41(1):5-9

Dalmau, J., et al. (2008). Anti-NMDA-receptor encephalitis: case series and analysis of the effects of antibodies. Lancet Neurol. 2008;7(12):1091-1098.

Daprati, E. et al., (2010). Body and movement: Consciousness in the parietal lobes. Neuropsychologia, 48, 756-762.

Dawbarn, D. & Allen, S. (2007). The Neurobiology of Alzheimer's Disease, Oxford University Press.

D'Cunha, T. M. et al. (2011). Oxytocin receptors in the nucleus accumbens shell are involved in the consolidation of maternal memory in postpartum rats, Hormones and behavior, 59, 14-21.

De Jong, L. W., et al. (2008), Strongly reduced volumes of putamen and thalamus in Alzheimer's disease: an MRI study Brain, Brain, 131 (12): 3277-3285

DeKeyser, R. M. (2001). The robustness of critical period effects in second language acquisition. Studies in Second Language Acquisition, 22, 499-533.

DeLong, M. R. & Wichmann, T. (2007). Circuits and circuit disorders of the basal ganglia, Archives of Neurology,64, 20-24.

Depue, B. E. et al., (2007), Prefrontal Regions Orchestrate Suppression of Emotional Memories via a Two-Phase Process, Science, 317, 215-219.

Desmurget, M., et al., (2009). Movement Intention After Parietal Cortex Stimulation in Humans, Science, 324, 811-813.

de Zubicaray GI, McMahon KL. (2009). Auditory context effects in picture naming investigated with event-related fMRI. Cogn Affect Behav Neurosci. 2009 Sep;9(3):260-9.

Downing PE, Jiang Y, Shuman M, Kanwisher N (2001) A cortical area selective for visual processing of the human body. Science 293:2470–2473.

DuBay, M. F., et al., (2011). Coping resources in individuals with aphasia, Aphasiology, 25, 1016-1029.

Dum, R. P., Strick, P. L. (2005). Frontal Lobe Inputs to the Digit Representations of the Motor Areas on the Lateral Surface of the Hemisphere The Journal of Neuroscience,25(6):1375-1386.

Duyckaerts, C. et al. (2009). Morphologic and molecular neuropathology of Alzheimer's disease, Ann Pharm Fr. 2009 Mar;67(2):127-35.

Eaton, E., et al., (2011). Mechanisms of change in the evolution of jargon aphasia, Aphasiology, 10.1080/02687038.2011.624584.

Edwards, M. et al. (2008). Parkinsons Disease and Other Movement Disorders, Oxford University Press.

Ehrlich, S. et al. (2010). The COMT Val108/158Met polymorphism and medial temporal lobe volumetry in patients with schizophrenia and healthy adults. Neuroimage 53, 992-1000.

Eichenbaum, H., & Lipton, P. A., (2008). Towards a functional organization of the medial temporal lobe memory system: Role of the parahippocampal and medial entorhinal cortical areas, Hippocampus, 18, 1314-1324.

Ellison-Wright, I., & Bullmore, B. (2009) Meta-analysis of diffusion tensor imaging studies in schizophrenia, Schizophrenia Research, 108, 3-10.

Endo, K. et al., (2010). Meaning, Phonological, Orthography and Kinestic Route of Reading and Writing: A Case with Alexia and Agraphia due to the Left Parietal Lesion. Brain Nerve. 2010 Sep;62(9):991-6.

Engel, T. et al., (2011). Expression of neurogenesis genes in human temporal lobe epilepsy with hippocampal sclerosis, Int J Physiol Pathophysiol Pharmacol 3, 38-47.

Esselly, R., et al. (2011). Handedness for grasping objects and pointing and the development of language in 14-month-old infants, Laterality: Asymmetries of Body, Brain and Cognition, 16, 565-585.

Ezzyat, Y., & Olson, I. R. (2008) The medial temporal lobe and visual working memory: Comparisons across tasks, delays, and visual similarity.Cognitive, Affective, & Behavioral Neuroscience, 8, 32-40.

Felician O, et al. (2009) Where is your shoulder? Neural correlates of localizing others' body parts. Neuropsychologia 47:1909–1916.

Ferguson, S. (2006). Temporal Lobe Epilepsy and the Mind-Brain Relationship: A New Perspective, Volume 76 (International Review of Neurobiology.Academic Press.

Fink, M., & Taylor, M.A,. (2009). The Catatonia Syndrome, Arch Gen Psychiatry. 66, 1173-1177.

Fotopoulou, A., et al. (2010). Implicit awareness in anosognosia for hemiplegia: unconscious interference without conscious re-representation. Brain, 133, 3564-3577.

Flamand-Roze, C., et al. (2011). Aphasia in border-zone infarcts has a specific initial pattern and good long-term prognosis, European Journal of Neurology, 18, 1397–1401.

Fleming, M. K., et al., (2010). Bilateral parietal cortex function during motor imagery. Exp. Brain, Res, 201, 499-508.

Fletcher, J. M. et al., (2006). Learning Disabilities: From Identification to Intervention, Guilford Press.

Fournier, N. M., et al., (2008). Impaired social cognition 30 years after hemispherectomy for intractable epilepsy: The importance of the right hemisphere in complex social functioning. Epilepsy & Behavior, 12, 460-471.

Fraser, J. A. et al., (2011) Disorders of the optic tract, radiation, and occipital lobe. Handb Clin Neurol. 102, 205-21.

Fridriksson, J., et al., (2009). Modulation of Frontal Lobe Speech Areas Associated With the Production and Perception of Speech Movements, Journal of Speech, Language, and Hearing Research Vol.52 812-819.

Fridriksson, J. et al., (2010). Activity in Preserved Left Hemisphere Regions Predicts Anomia Severity in Aphasia, Cereb. Cortex (2010) 20 (5): 1013-1019.

Fridriksson, J., et al., (2010). Impaired Speech Repetition and Left Parietal Lobe Damage. J. Neuroscience, 30, 11057-11061.

Fuster, J. (2008) The Prefrontal Cortex. Academic Press.

Gazzaniga, M.S., (2000). Cerebral specialization and interhemispheric communication -- does the corpus callosum enable the human condition? Brain 123, 1293–1326.

Gentile, G., et al., (2010). Integration of Visual and Tactile Signals From the Hand in the Human

Brain: An fMRI Study. Journal of Neurophysiology, 105, 910-922.

Geuze, E. et al. (2008). Reduced GABAA benzodiazepine receptor binding in veterans with post-traumatic stress disorder, Molecular Psychiatry 13, 74–83.

Glasser, M. F. & Rilling, J. K. (2008). DTI Tractography of the Human Brain's Language Pathways, Cortex, 18, 2471-482.

Goldenberg, G. (2009). Apraxia and the parietal lobes. Neuropsychologia, 47, 1449-1459

González, J, & McLennan, C. T. (2009). Hemispheric Differences in the Recognition of Environmental Sounds. Psychological Science July 2009 vol. 20 no. 7 887-894.

Gottlieb, J., & Snynder, L. H. (2010). Spatial and non-spatial functions of the parietal cortex. Current Opinion in Neurobiology, 20, 731-740.

Grahn, J. A. et al. (2009). The role of the basal ganglia in learning and memory: neuropsychological studies, Behavioural Brain Research, 199, 53-60.

Greenlee, J. D., et al. (2004). A functional connection between inferior frontal gyrus and orofacial motor cortex in human, J Neurophysiol. 2004 Aug;92(2):1153-64.

Grossman, M. et al. (2003) Neural basis for semantic memory diY- culty in Alzheimer's disease: an fMRI study. Brain 126(Pt 2):292–311

Grueter, B. A. et al. (2011). Integrating synaptic plasticity and striatal circuit function in addiction, Current Opinion in Neurobiology, doi:10.1016/j.conb.2011.09.009

Haber, S. N. (2009). The cortico-basal ganglia integrative network: the role of the thalamus, Brain Research Bulletin, 78, 69-74.

Halbach, O. V. B. U. & Demietzel, R. (2006). Neurotransmitters and Neuromodulators: Handbook of Receptors and Biological Effects, Wiley.

Hama, S., et al., (2007). Post-stroke affective or apathetic depression and lesion location: left frontal lobe and bilateral basal ganglia, Eup, Arch. Psych. Clinical Neuroscience, 257, 149-152.

Hampshire, A., et al. (2010). The role of the right inferior frontal gyrus: inhibition and attentional control. NeuroImage, 50, 1313-1319.

Harciarek, M., et al., (2006). Defective comprehension of emotional faces and prosody as a result of right hemisphere stroke: Modality versus emotion-type specificity Journal of the International Neuropsychological Society, 12, 774-781.

Hashimoto, R., Sakai, K.L., (2002). Specialization in the left prefrontal cortex for sentence comprehension. Neuron 35, 589–597.

Hegde, J., & Felleman, D. J. (2007). Reappraising the Functional Implications of the Primate Visual Anatomical Hierarchy. The Neuroscientist, 13, 416-421.

Heim, C. et al. (2008). The dexamethasone/corticotropin-releasing factor test in men with major depression: role of childhood trauma, Biological Psychiatry, 63, 398-405.

Heilman, K. M. (2008), Right Hemispheric Neurobehavioral Syndromes, In Stein, J. et al. (Eds). Stroke Recovery and Rehabilitation, Demos Medical.

Helm-Estabrooks, N. (2011). Treating Attention To Improve Auditory Comprehension Deficits Associated With Aphasia, Perspectives on Neurophysiology and Neurogenic Speech and Language Disorders ,21 64-71.

Henry, T. R. et al. (2011). Hippocampal Sclerosis in Temporal Lobe Epilepsy: Findings at 7 T, Radiology, 261, 199-209.

Heydrich, L., et al., (2011). Partial and full own-body illusions of epileptic origin in a child with right temporoparietal epilepsy Epilepsy & Behavior, 20, 583-586.

Hobson, J. A. (2004). Dreaming: An Introduction to the Science of Sleep. Oxford University Press.

Hoeft, F. et al. (2007) Functional and morphometric brain dissociation between dyslexia and reading ability. Proc. Natl. Acad. Sci. USA, 104 4234–4239.

Holmes, C. et al. (2008). Long-term effects of Abeta42 immunisation in Alzheimer's disease: follow-up of a randomised, placebo-controlled phase I trial. Lancet 372 (9634): 216–23.

Holstege, G. & Huynh, H. H. (2011). Brain circuits for mating behavior in cats and brain activations and de-activations during sexual stimulation and ejaculation and orgasm in humans. Hormones and Behavior. 59, 702-707.

Honeycutt, J. M., et al., (2011). Developmental Implications of Mental Imagery in Childhood Imaginary Companions, Imagination, Cognition and Personality, 31, 79-98.

Horwitz B., et al. (2003). Activation of Broca's area during the production of spoken and signed language: a combined cytoarchitectonic mapping and PET analysis.Neuropsychologia. 2003;41(14):1868-76.

Hugdahl, K., et al. (2007). Auditory hallucinations in schizophrenia: the role of cognitive, brain structural and genetic disturbances in the left temporal lobe, Front Hum Neurosci. 1, 6.

Hugdahl, K., et al., (2009). Left temporal lobe structural and functional abnormality underlying auditory hallucinations in schizophrenia. Front Neurosci. 3, 34–45.

Hurlock, E. B.; Burstein, M. (1932). The imaginary playmate: a questionnaire study. The Pedagogical Seminary and Journal of Genetic Psychology, Vol 41, 1932, 380-392.

lmeida, L. G. et al., (2010). Reduced right frontal cortical thickness in children, adolescents and adults with ADHD and its correlation to clinical variables: A cross-sectional study Journal of Psychiatric Research, 44, 214-1223.

Iqbal, K., et al. (2005). Tau pathology in Alzheimer disease and other tauopathies. Biochim Biophys Acta 1739 (2–3): 198–210.

Ito, M. (2011). The Cerebellum: Brain for an Implicit Self, FT Press.

Iversen, L., et al. (2009). Dopamine Handbook, Oxford University Press.

Jackson, M. C., et al., (2008). Neural Correlates of Enhanced Visual Short-Term Memory for Angry Faces: An fMRI Study. PLoS ONE 3(10): e3536. doi:10.1371/journal.pone.0003536

Jerde, T. A. et al. (2011). Dissociable systems of working memory for rhythm and melody.Neuroimage. 2011 Aug 15;57(4):1572-9.

Johnson MK, et al, (2005). Using fMRI to investigate a component process of reflection: prefrontal correlates of refreshing a just-activated representation. Cogn. Affect. Behav. Neurosci. 5:339–61.

Joseph, R. (1977). Females are more aroused by attractive women versus attractive men as based on pupil dilation and viewing time. Unpublished research and manuscript (deemed to "politically incorrect for the 1970s).

Joseph, R., Hess, S., & Birecree, E. (1978). Effects of sex hormone manipulations on exploration and sex differences in maze learning. Behavioral Biology, 24, 364-377,

Joseph, R., & Casagrande, V. A. (1978). Visual field defects and morphological changes resulting from monocular deprivation in primates. Proceedings of the Society for Neuroscience, 4, 1982, 2021.

Joseph, R. (1979). Effects of rearing environment and sex on learning, memory, and competitive exploration. Journal of Psychology, 101, 37-43.

Joseph, R. & Casagrande, V. A. (1978). Visual field defects and recovery following lid closure in a prosimian primate. Behavioral Brain Research, 1, 150-178.

Joseph, R., & Gallagher, R. E. (1980). Gender and early environmental influences on learning, memory, activity, overresponsiveness, and exploration. journal of Developmental Psychobiology, 13, 527-544.

Joseph, R. (1980). Awareness, the origin of thought, and the role of conscious self-deception in resistance and repression. Psychological Reports, 46, 767-781.

Joseph, R., Forrest, N., Fiducia, N., Como, P., & Siegel, J. (1980). Electrophysiological and behavioral correlates of arousal. Physiological Psychology, 9, 90-95.

Joseph, R. (1982). The Neuropsychology of Development. Hemispheric Laterality, Limbic Language, the Origin of Thought. Journal of Clinical Psychology, 44 4-33.

Joseph, R. (1984). La Neuropsiciologia Del Desarrollo: Lateralidad, Hemisferica, Lenguaje Limbico, y el Origen del pensamiento. Archives of Psychiatry and Neurology, Venezolanos, 30, 25-52.

Joseph, R., Gallagher, R., E., Holloway, J., & Kahn, J. (1984). Two brains, one child: Interhemispheric transfer and confabulation in children aged 4, 7, 10. Cortex, 20, 317-331.

Joseph, R., & Gallagher, R. E. (1985). Interhemispheric transfer and the completion of reversible operations in non-conserving children. Journal of Clinical Psychology, 41, 796-800.

Joseph, R. (1985). Competition between women. Psychology, 22, 1-11.

Joseph, R. (1986). Reversal of language and emotion in a corpus callosotomy patient. Journal of Neurology, Neurosurgery, & Psychiatry, 49, 628-634.

Joseph, R. (1986). Confabulation and delusional denial: Frontal lobe and lateralized influences. Journal of Clinical Psychology, 42, 845-860.

Joseph, R. (1988) The Right Cerebral Hemisphere: Emotion, Music, Visual-Spatial Skills, Body Image, Dreams, and Awareness. Journal of Clinical Psychology, 44, 630-673.

Joseph, R. (1988). Dual mental functioning in a split-brain patient. Journal of Clinical Psychology, 44, 770-779.

Joseph, R. (1990). Neuropsychology, Neuropsychiatry, Behavioral Neurology, Plenum, New York.

Joseph, R. (1990). The right cerebral hemisphere. Emotion, music, visual-spatial skills, body image, dreams, awareness. In A. E. Puente and C. R. Reynolds (series editors). Critical Issues in Neuropsychology. Neuropsychology, Neuropsychiatry, Behavioral Neurology. Plenum, New York.

Joseph, R. (1990). The left cerebral hemisphere. Aphasia, alexia, agraphia, agnosia, apraxia, language and thought. In A. E. Puente and C. R. Reynolds (series editors). Critical Issues in Neuropsychology. Neuropsychology, Neuropsychiatry, Behavioral Neurology. Plenum, New York.

Joseph, R. (1990). The limbic system. Emotion, laterality, unconscious mind. In A. E. Puente and C. R. Reynolds (series editors). Critical Issues in Neuropsychology. Neuropsychology, Neuropsychiatry, Behavioral Neurology. Plenum, New York.

Joseph, R. (1990). The Frontal Lobes. In A. E. Puente and C. R. Reynolds (series editors). Critical Issues in Neuropsychology. Neuropsychology, Neuropsychiatry, Behavioral Neurology. Plenum, New York.

Joseph, R. (1990). The Parietal Lobes. In A. E. Puente and C. R. Reynolds (series editors). Critical Issues in Neuropsychology. Neuropsychology, Neuropsychiatry, Behavioral Neurology. Plenum, New York.

Joseph, R. (1990). The Temporal Lobes. In A. E. Puente and C. R. Reynolds (series editors). Critical Issues in Neuropsychology. Neuropsychology, Neuropsychiatry, Behavioral Neurology. Plenum, New York.

Joseph, R. (1990). The Occipital Lobes. In A. E. Puente and C. R. Reynolds (series editors). Critical Issues in Neuropsychology. Neuropsychology, Neuropsychiatry, Behavioral Neurology. Plenum, New York.

Joseph, R. (1990). Cerebral and cranial trauma. In A. E. Puente & C. R. Reynolds (series editors). Critical Issues in Neuropsychology. Neuropsychology, Neuropsychiatry, Behavioral Neurology. Plenum, New York.

Joseph, R. (1990). Stroke and cerebrovascular disease. In A. E. Puente and C. R. Reynolds (series editors). Critical Issues in Neuropsychology. Neuropsychology, Neuropsychiatry, Behavioral Neurology. Plenum, New York.

Joseph, R. (1990). Cerebral neoplasms. In A. E. Puente and C. R. Reynolds (series editors). Critical Issues in Neuropsychology. Neuropsychology, Neuropsychiatry, Behavioral Neurology. Plenum, New York.

Joseph, R. (1992) The Limbic System: Emotion, Laterality, and Unconscious Mind. The Psychoanalytic Review, 79, 405-456.

Joseph, R. (1992). The Right Brain and the Unconscious. New York, Plenum.

Joseph, R. (1993). The Naked Neuron: Evolution and the Languages of the Body and Brain.

New York, Plenum Press.

Joseph, R. (1994) The limbic system and the foundations of emotional experience. In V. S. Ramachandran (Ed). Encyclopedia of Human Behavior. San Diego, Academic Press.

Joseph, R. (1996). Neuropsychiatry, Neuropsychology, Clinical Neuroscience, 2nd Edition. 21 chapters, 864 pages. Williams & Wilkins, Baltimore.

Joseph, R. (1998). The limbic system. In H.S. Friedman (ed.), Encyclopedia of Human health, Academic Press. San Diego.

Joseph, R. (1998). Traumatic amnesia, repression, and hippocampal injury due to corticosteroid and enkephalin secretion. Child Psychiatry and Human Development. 29, 169-186.

Joseph, R. (1998). Flavonoid substance and/or flavone glycosides substance as a treatment for disorders of the brain. United States Department of Commerce: Patent & Trademark Office, March, # 60/080,768.

Joseph, R. (1998). Combined use of Ginko Biloba and Hypericum Perforatum (Saint John's Wort) as a treatment for disorders of the brain. United States Department of Commerce: Patent & Trademark Office, March, # 60/080,769.

Joseph, R. (1998). Olfactory substance and stem cells as a treatment for disorders of the brain. United States Department of Commerce: Patent & Trademark Office, March, # 60/080,770.

Joseph, R. (1999). Frontal lobe psychopathology: Mania, depression, aphasia, confabulation, catatonia, perseveration, obsessive compulsions, schizophrenia. journal of Psychiatry, 62, 138-172.

Joseph, R. (1999). Environmental influences on neural plasticity, the limbic system, and emotional development and attachment, Child Psychiatry and Human Development. 29, 187-203.

Joseph, R. (1999). The neurology of traumatic "dissociative" amnesia. Commentary and literature review. Child Abuse & Neglect. 23, 715-727

Joseph, R. (2000). Astrobiology, the Origin of Life, and the Death of Darwinism. University Press California.

Joseph, R. (2000). Female Sexuality: The Naked Truth. University Press California.

Joseph, R. (2000). Limbic language/language axis theory of speech. Behavioral and Brain Sciences. 23, 439-441.

Joseph, R. (2000). Fetal brain behavioral cognitive development. Developmental Review, 20, 81-98.

Joseph, R. (2000). The evolution of sex differences in language, sexuality, and visual spatial skills. Archives of Sexual Behavior, 29, 35-66.

Joseph, R. (2001). Biological Substances to Induce Sexual Arousal and as a Treatment for Sexual Dysfunction. United States Department of Commerce: Patent & Trademark Office, January 12, 2001 #60/260,910.

Joseph, R. (2001). Biological Substances to Induce Sexual Arousal, Sexual Behavior, Ovulation, Pregnancy, and Treatment for Sexual Dysfunction. United States Department of Commerce: Patent & Trademark Office, February # 60/.

Joseph, R. (2001). Clinical Neuroscience, 34 chapters, 1,500 pages. Academic Press.

Joseph, R. (2001). The Limbic System and the Soul: Evolution and the Neuroanatomy of Religious Experience. Zygon, the Journal of Religion & Science, 36, 105-136.

Joseph, R. (2002). Biological Substances to Induce Sexual Arousal and as a Treatment for Sexual Dysfunction. Patent Pending: United States Department of Commerce: Patent & Trademark Office, February, 2002 #10/047,906

Joseph, R. (2002/2003). NeuroTheology: Brain, Science, Spirituality, Religious Experience. University Press.

Joseph, R. (2003). Emotional Trauma and Childhood Amnesia. journal of Consciousness & Emotion, 4, 151-178.

Joseph, R. (2010). Extinction, Metamorphosis, Evolutionary Apoptosis, and Genetically Programmed Species Mass Death. In "The Biological Big Bang," Edited by Chandra Wickramasinghe, Science Publishers, Cambridge, MA

Joseph, R. (2010). Sex on Mars In: "The Human Mission to Mars: Colonizing the Red Planet", Edited by Joel Levine, Ph.D., Journal of Cosmology, 12, 4034-4050.

Joseph, R. (2011). Quantum Physics and the Multiplicity of Mind: Split-Brains, Fragmented Minds, Dissociation, Quantum Consciousness. "The Universe and Consciousness", Edited by Sir Roger Penrose, FRS, Ph.D., & Stuart Hameroff, Ph.D. Science Publishers, Cambridge, MA.

Joseph, R. (2011). Neuroanatomy of Free Will: Loss of Will, Against the Will, Alien Hand. "The Universe and Consciousness", Edited by Sir Roger Penrose, FRS, Ph.D., & Stuart Hameroff, Ph.D. Science Publishers, Cambridge, MA.

Joseph, R. (2011). The Split Brain: Two Brains - Two Minds "The Universe and Consciousness", Edited by Sir Roger Penrose, FRS, Ph.D., & Stuart Hameroff, Ph.D. Science Publishers, Cambridge, MA.

Joseph, R. (2011). Origins of Thought: Consciousness, Language, Egocentric Speech and the Multiplicity of Mind "The Universe and Consciousness", Edited by Sir Roger Penrose, FRS, Ph.D., & Stuart Hameroff, Ph.D. Science Publishers, Cambridge, MA.

Joseph, R. (2011). Dreams and Hallucinations: Lifting the Veil to Multiple Perceptual Realities "The Universe and Consciousness", Edited by Sir Roger Penrose, FRS, Ph.D., & Stuart Hameroff, Ph.D. Science Publishers, Cambridge, MA.

Joseph, R. (2011). Evolution of Paleolithic Cosmology and Spiritual Consciousness, and the Temporal and Frontal Lobes Journal of Cosmology, 2011, Vol. 14.

Kaan, E., & Swaab, T. Y. (2002). The brain circuitry of syntactic comprehension. Trends in Cognitive Sciences, 6, 350-356.

Kadosh, R. C. & Walsh, V. (2009). Numerical representation in the parietal lobes: Abstract or not abstract? Behavioral and Brain Sciences, 32, 313-328.

Kapogiannis, D. et al., (2009). Cognitive and neural foundations of religious belief. PNAS, 106, 4876-4881.

Karnath HO, Baier B, Nägele T (2005) Awareness of the functioning of one's own limbs mediated by the insular cortex? J Neurosci 25:7134–7138

Keenan JP, Nelson A, O'Connor M, Pascual-Leone A (2001) Neurology: Self-recognition and the right hemisphere. Nature 409: 305.

Keller, T. A., Carpenter, P. A., & Just, M. A. (2001). The neural bases of sentence comprehension: A fMRI examination of syntactic and lexical processing. Cerebral Cortex, 11, 223-237.

Kerfott, E. C., & Williams, C. L. (2011). Interactions between brainstem noradrenergic neurons and the nucleus accumbens shell in modulating memory for emotionally arousing events, Learn. Mem. 18: 405-413.

Kidd, E. et al., (2010). The personality correlates of adults who had imaginary companions in childhood. Psychological Reports, 107, 163-172.

Kim, Y.-D., et al., (2010). Callosal alien hand sign following a right parietal lobe infarction. Journal of Clinical Neuroscience, 17, 796-797.

Kleinschmidt, A. (2011). Gerstmann Meets Geschwind: A Crossing (or Kissing) Variant of a Subcortical Disconnection Syndrome? The Neuroscientist, 10, 13.

Kloesel, B., et. al., (2010). Sequelae of a left-sided parietal stroke: Posterior alien hand syndrome. Neurocase, 16, 488-493.

Klostermann, E. C. et al., (2009). Activation of right parietal cortex during memory retrieval of nonlinguistic auditory stimuli, Cognitivem, Affective & Behavioral Neuroscience, 9, 242-248.

Konen, C. S. & Kastner, S. (2008). Two hierarchically organized neural systems for object information in human visual cortex. Nature Neuroscience, 11, 224-231.

Koss S, et al. (2010). Numerosity impairment in corticobasal syndrome. Neuropsychology, 24:476-492.

Kriegeskorte, N. et al., (2008). Matching categorical object representations in inferior temporal cortex of man and monkey Neuron, 60, 1126-1141.

Kronbichler, M. et al (2007). Developmental dyslexia: Gray matter abnormalities in the occipitotemporal cortex, Human Brain Mapping, 29, 613-625.

Lader, M. et al. (2010). Sleep and Sleep Disorders, Springer.

Lane, Z. P. et al., (2011). Differentiating Psychosis versus Fluent Aphasia. Clinical Schizophrenia & Related Psychoses, 4, 258-261.

Lanyon, L. J., et al., (2009). Disconnection of cortical face network in prosopagnosia revealed by diffusion tensor imaging. Journal of Vision, 9, 482.

LaPlante, E. (2000). Seized: Temporal Lobe Epilepsy as a Medical, Historical, and Artistic Phenomenon, iUniverse

Leonard, C. M., et al., (2008). Size Matters: Cerebral Volume Influences Sex Differences in Neuroanatomy, Cerebral Cortex, 18, 2920-2931.

Lever, C. et al. (2010). Environmental novelty elicits a later theta phase of firing in CA1 but not subiculum, Hippocampus, 20, 229-234.

Levins, S. M., et al., (2011). Role of the left amygdala and right orbital frontal cortex in emotional interference resolution facilitation in working memory. Neuropsychologia, 49, 3201-3212.

Lindell, A.K., (2006). In your right mind: right hemisphere contributions to language processing and production. Neuropsychol. Rev. 16, 131–148.

Lindenberg, R., & Scheef, L. (2007). Supramodal language comprehension: Role of the left temporal lobe for listening and reading. Neuropsychologia, 45, 2407-2415.

Links, P., et al. (2010). Training verb and sentence production in agrammatic Broca's aphasia, Aphasiology, 24, 11, 1303-1325.

Linnman, C et al. (2011). Traumatic stress activates the amygdala, putamen, and caudate, and a disconnection between the amydala-putamen in those with PTSD under conditions of additional stress. Biology of Mood & Anxiety Disorders 2011, 1:8.

Lintas, A. et al. (2011). Identification of a Dopamine Receptor-Mediated Opiate Reward Memory Switch in the Basolateral Amygdala–Nucleus Accumbens Circuit, J. Neuroscience, 31, 11172-11183.

Liu, J., et al., (2010). Perception of Face Parts and Face Configurations: An fMRI Study. Journal of Cognitive Neuroscience, 22, 203-211.

Luef, G. J. (2008). Epilepsy and sexuality. Seizure, 17, 127-130.

Maldonado, L., et al., (2011). Surgery for gliomas involving the left inferior parietal lobule: new insights into the functional anatomy provided by stimulation mapping in awake patients, Journal of Neurosurgery, 115 / No. 4 / 770-779.

Madden, M. & Sunda, T. (2009). Beyond hippocampal sclerosis. The rewired hippocampus in temporal lobe epilepsy Neurology,

Mallick, B. N. et al. (2011). Rapid Eye Movement Sleep: Regulation and Function, Cambridge University Press.

Malikovic, A. et al., (2011). Occipital sulci of the human brain: variability and morphometry. Anatomical Science, s12565-011-0118-6.

Malloy, C. et al., (2011). Is Traumatic Brain Injury A Risk Factor for Schizophrenia? A Meta-Analysis of Case-Controlled Population-Based Studies. Schizophr Bull . 37, 1104-1110

Manoach DS, Greve DN, Lindgren KA, Dale AM. (2003). Identifying regional activity associated with temporally separated components of working memory using event-related functional MRI. NeuroImage 20(3):1670–84.

Medland, S., et al. (2009). Genetic influences on handedness: Data from 25,732 Australian and Dutch twin families, Neuropsychologia, 47, 330-337.

Menghini, D. et al. (2008). Structural Correlates of Implicit Learning Deficits in Subjects with Developmental Dyslexia, Annals of the New York Academy of Sciences, 1145, 212–221.

Mesulam, M., et al. (2009). Neurology of anomia in the semantic variant of primary progressive aphasia, Brain 132 (9): 2553-2565.

Miceli, G. et al., (2008). Acute auditory agnosia as the presenting hearing disorder in males.

Neurological Sciences, 29, 459-462.

Miller, B. L., & Cummings, J. L. (2006). The Human Frontal Lobes. Guilford Press.

Millonig, A., et al. (2011). Supernumerary phantom limb as a rare symptom of epileptic seizures—case report and literature review. Epilepsia, 52, 1528-1167.

Montemurro, M. A., et al., (2008). Phase-of-Firing Coding of Natural Visual Stimuli in Primary Visual Cortex. Current Biology, 18, 375-380.

Mormann, F. et al. (2008). Independent Delta/Theta Rhythms in the Human Hippocampus and Entorhinal Cortex, Front Hum Neurosci. 10.3389/neuro.09.003.2008.

Morris JP, Pelphrey KA, McCarthy G (2006) Occipitotemporal activation evoked by the perception of human bodies is modulated by the presence or absence of the face. Neuropsychologia 44:1919–1927.

Moser, D., et al., (2009). Temporal Order Processing of Syllables in the Left Parietal Lobe. J. Neuroscience, 29, 12568-12573.

Mulroy, N., et al. (2011). Alexia without Agraphia. The Irish Medical Journal, 104, 9.

Murray, E. A. et al., (2007). Visual Perception and Memory: A New View of Medial Temporal Lobe Function in Primates and Rodents. Annual Reviews Neuroscience, 30: 99-122.

Mzoyer, B., et al. (2009). Brain, language, and handedness: a family affair, Nature Precedings, hdl:10101/npre.2009.2982.1

Nachev, P., Kennard, C., & Husain, M. (2008). Functional role of the supplementary and pre-supplementary motor areas Nature Reviews Neuroscience 9, 856-869.

Nasr, S. & Tootell, R. (2011). Contribution of anterior temporal lobe in recognition of face and non-face objects. Journal of Vision, 11, #69.

Nassi, N. J., & Callaway, E. M. (2009). Parallel Processing Strategies of the Primate Visual System. Nat Rev Neurosci. 10 360–372.

Nelson, S. M. et al., (2010). A Parcellation Scheme for Human Left Lateral Parietal Cortex. Neuron, 67, 156-170.

Newman, S. D., Just, M. A., & Carpenter, P. A. (2002). Synchronization of the human cortical working memory network. NeuroImage, 15, 810-822.

Nico, D. et al., (2010). The role of the right parietal lobe in anorexia nervosa. Psychological Medicine, 40, 1531-1539.

Niki, C., et al., (2009). Disinhibition of sequential actions following right frontal lobe damage, Cognitive Neuropsychology, 26, 266-285.

Norton, A. et al., (2009). Melodic Intonation Therapy. Annals of the New York Academy of Sciences, 1169, 4310-436.

Ogar, J. M. et al., (2011). Semantic dementia and persisting Wernicke's aphasia: Linguistic and anatomical profiles. Brain & Language, 117, 28-33.

Olson, I. R. & Berryhill, M. (2009). Some surprising findings on the involvement of the parietal lobe in human memory. Neurobiology of Learning and Memory, 91, 155-165.

Özdemi, E., et al., (2006). Shared and distinct neural correlates of singing and speaking, NeuroImage, 33, 626-635.

Pai, A. R., et al. (2011). Global aphasia without hemiparesis: A case series, Ann Indian Acad Neurol. 2011 Jul-Sep; 14(3): 185–188.

Pantelyat,A., et al., (2011). Acalculia in Autopsy-Proven Corticobasal Degeneration. Neurology, 76, S61-S62.

Papoutsi M., et al., (2009). From phonemes to articulatory codes: an fMRI study of the role of Broca's area in speech production.Cereb Cortex. 2009 Sep;19(9):2156-65.

Partonen. T., & Pandi-Perumal, S. R. (2010). Seasonal affective disorder: practice and research. Oxford University Press.

Paulesu, E., et al. (2000). A cultural effect on brain function. Nat. Neurosci. 3, 91–96.

Peelen MV, Downing PE (2007) The neural basis of visual body perception. Nat Rev Neurosci 8:636–648

Perani, D., et al. (1998). The bilingual brain — proficiency and age of acquisition of the second language. Brain 121, 1841–1852.

Peretz, I. et al., (2009). Music Lexical Networks. Annals of the New York Academy of Sciences, 1169, 256–265.

Perez-Costas, E. et al. (2010). Basal ganglia pathology in schizophrenia: dopamine connections and anomalies, Journal of Neurochemistry, 113, 287-302.

Persaud, N. et al. (2011).Awareness-related activity in prefrontal and parietal cortices in blind-sight reflects more than superior visual performance.

Philippi, C. L., et al., (2009). Damage to Association Fiber Tracts Impairs Recognition of the Facial Expression of Emotion, Journal of Neuroscience, 29, 15089-15099

Piller, R. (2009). Cerebral specialization during lucid dreaming: A right hemisphere hypothesis. Dreaming, 19, 273-286.

Platek SM, Keenan JP, Gallup GG, Mohamed FB (2004) Where am I? The neurological correlates of self and other. Cog Brain Res 19: 114–122.

Pisella, L., et al., (2010). Right-hemispheric dominance for visual remapping in humans, Phil. Trans. R. Soc. B, 1564, 572-585.

Pitcher, D. et al.. (2011). The role of the occipital face area in the cortical face perception network. Exp. Brain Res. 209, 481-493.

Plessen, K. J. et al. (2006). Hippocampus and Amygdala Morphology in Attention-Deficit/Hyperactivity Disorder, Arch Gen Psychiatry. 63:795-807.

Poeppel, D., (2003). The analysis of speech in different temporal integration windows: cerebral lateralization as "asymmetric sample in time". Speech Commun. 41, 245–255.

Postle BR. (2006). Working memory as an emergent property of the mind and brain. Neuroscience 139:23–38.

Preston, A. R. et al. (2010). High-resolution fMRI of content-sensitive subsequent memory responses in human medial temporal lobe, J. of Cognitive Neuroscience, 22, 156-173.

Priebe, N. J. & Ferster, D. (2008). Inhibition, Spike Threshold, and Stimulus Selectivity in Primary Visual Cortex. Neuron, 57, 482-497.

Prigatano, G. P., & Wolf, T. R. (2010) Anton's Syndrome and Unawareness of Partial or complete blindness. In Prigatano, G. P. (ed), The Study of Agnosognosia. Oxford University Press.

Prigatano, G. P. (2010), The Study of Agnosognosia. Oxford University Press.

Quinkert, A. W. et al. (2010). Temporal patterning of pulses during deep brain stimulation affects central nervous system arousal, Behavioural Brain Research, 214, 377-385.

Quiroga, R. W. et al., (2008). Sparse but not [] Grandmother-cell'coding in the medial temporal lobe. Trends in Cognitive Sciences,12, 87091.

Ramayya, A. G., et al., (2010). A DTI Investigation of Neural Substrates Supporting Tool Use. Cerebral Cortex, 20, 507-516.

Ranganath C, D'Esposito M. (2005). Directing the mind's eye: prefrontal, inferior and medial temporal mechanisms for visual working memory. Curr. Opin. Neurobiol. 15:175–82.

Rauschecker, J. P. & Scott, S. K. (2009). Maps and streams in the auditory cortex: nonhuman primates illuminate human speech processing. Nat. Neuroscience, 12, 718-724.

Reed, J. L. et al. (2008). Widespread spatial integration in primary somatosensory cortex, PNAS, 105, 10233-10237

Reuter-Lorenz PA, Jonides J. (2007). The executive is central to working memory: insights from age, performance and task variations. In Variations in Working Memory, ed. AR Conway, C Jarrold, MJ Kane, A Miyake, JN Towse, pp. 250–70. London/New York: Oxford Univ. Press

Riecker, A., (2000). Articulatory/phonetic sequencing at the level of the anterior perisylvian cortex: a functional magnetic resonance imaging (fMRI) study. Brain Lang. 75, 259–276.

Rilling, J. F. et al., (2008). The evolution of the arcuate fasciculus revealed with comparative DTI, Nature Neuroscience, 11, 426-428.

Risberg J., Grafman, J. (2006). The Frontal Lobes: Development, Function and Pathology, Cambridge University Press

Rizzolatti, G., Craighero, L. (2004). TRhe mirror-neuron system, Ann. Rev. Neuros. 27, 169–92.

Rizzolatti G, Fadiga L, Fogassi L, Gallese V. (1996). Premotor cortex and the recognition of motor actions. Cogn. Brain Res. 3:131– 141.

Rizzolatti G, Luppino G, Matelli M. (1998). The organization of the cortical motor system: new concepts. Electroencephalogr. Clin. Neurophysiol. '06:283–296.

Rizzolatti, G., Fabbri-Destro, M., Cattaneo, L. (2009). Mirror neurons and their clinical relevance. Nat Clin Pract Neurol 5 (1): 24–34.

Roether, C. L., et al., (2008). Lateral asymmetry of bodily emotion expression, Current Biology, 18, R329-R330.

Rogalski, E. J., & Mesulam, M. M. (2009). Clinical Trajectories and Biological Features of Primary Progressive Aphasia (PPA). Curr Alzheimer Res. 2009 August; 6(4): 331–336.

Roskies, A.L., et al, (2001). Task-dependent modulation of regions in the left inferior frontal cortex during semantic processing. J. Cogn. Neurosci. 13, 829–843.

Ross, E. D., & Monnot, M., (2008), Neurology of affective prosody and its functional–anatomic organization in right hemisphere, Brain and Language, 104, 51-74.

Rueckert, L., & Naybar, N. (2008). Gender differences in empathy: The role of the right hemisphere, Brain and Cognition, 67, 162-167.

Rusconi, E., & Kleinschmid, A (2011). Gerstmann's syndrome: where does it come from and what does that tell us? Future Neurology 6, 23-32.

Russell, C. et al. (2010). A deficit of spatial remapping in constructional apraxia after right-hemisphere stroke, Brain, 133, 1239-1251

Saur, D. et al;., (2008). Ventral and dorsal pathways for language. PNAS, 105, 18035-18040.

Schapira, A. H. V., et al. (2010). Movement Disorders, Saunders.

Scheuerecker, J. et al. (2009). Cerebral network deficits in post-acute catatonic schizophrenic patients measured by fMRI, Journal of Psychiatric Research, 43, 607-614.

Schirmer, A., & Kotz, S. A. (2006). Beyond the right hemisphere: brain mechanisms mediating vocal emotional processing. Trends in Cognitive Sciencesm 10, 24-30.

Schlagga, B.L. & McCandliss, B.D. (2007). Development of neural systems for reading. Annu. Rev. Neurosci., 30 475–503.

Schmahmann, J. M. (1997). The Cerebellum and Cognition, Academic Press.

Schwarzlose, R. F. et al., (2008). The distribution of category and location information across object-selective regions in human visual cortex. PNAS, 105, 4447-4452.

Selvage, D. L., et al. (2006). Role played by brainstem neurons in regulating testosterone secretion via a direct neural pathway between the hypothalamus and the testes, Endocrinology, 147, 3070-3075.

Seubert, J., et al., (2008). Straight after the turn: The role of the parietal lobes in egocentric space processing. Neurocase, 14, 204-219.

Shahana, N. et al. (2011). Neurochemical alteration in the caudate: Implications for the pathophysiology of bipolar disorder. Psychiatry Res. 193, 107-112.

Shinoura, N., et al., (2010). Upper Portion of Area 19 and the Deep White Matter in the Left Inferior Parietal Lobe, Including the Superior Longitudinal Fasciculus, Results in Alexia with Agraphia, European Neurology, 64, 224-229.

Shinoura, N. et al., (2010). Damage of left temporal lobe resulting in conversion of speech to Sutra, a Buddhist prayer stored in the right hemisphere. Neurocase, 16, 317-320.

Shomstein, S., et al., (2010). Top-down and bottom-up attentional guidance: investigating the role of the dorsal and ventral parietal cortices. Exp. Brain Res. 206. 197-208.

Shulman, G. L., et al. (2010). Right Hemisphere Dominance during Spatial Selective Attention and Target Detection Occurs Outside the Dorsal Frontoparietal Network. The Journal of Neuroscience, 30, 3640-3651.

Siddiqui, A. & Shaharyar, S. (2006). Role of 5-ht7 receptors on lordosis behaviour and lh release in the female rats, Pak J Physiol, 2,1.

Siegel, A. M., et al., (2003). The Parietal Lobe, Lippincott Williams & Wilkins.

Simons, J. S. et al., (2008). Processes subserved by the human parietal lobe appear to be recruited to support memory function, Neuropsychologia, 46, 1185-1191.

Simon, L. et al., (2002). Topographical Layout of Hand, Eye, Calculation, and Language-Related Areas in the Human Parietal Lobe, Neuron, 33, 475-487.

Simonyan, K., et al., (2007). Functional neuroanatomy of human voluntary cough and sniff production. Neuroimage. 2007 Aug 15;37(2):401-9.

Singh-Curry, V. & Husain, M. (2009). The functional role of the inferior parietal lobe in the dorsal and ventral stream dichotomy. Neuropsychologia, 47, 1434-1448.

Smania, N. et al., (2010). How Long Is the Recovery of Global Aphasia? Twenty-Five Years of Follow-up in a Patient With Left Hemisphere Stroke, Neurorehabil Neural Repair November/ December 2010 vol. 24 no. 9 871-875.

Smith , E., and Delargy, M. (2005). Locked-in syndrome, BMJ. 330, 406-408.

Snyder, K., et al. (2011). Corticotropin-Releasing Factor in the Norepinephrine Nucleus, Locus Coeruleus, Facilitates Behavioral Flexibility, Neuropsychopharmacology, doi:10.1038/ npp.2011.218.

Sowell, E. R. et al., (2007). Sex Differences in Cortical Thickness Mapped in 176 Healthy Individuals between 7 and 87 Years of Age. Cerebral Cortex, 17, 1550-1560.

Spengler S, von Cramon DY, Brass M (2009) Was it me or was it you? How the sense of agency originates from ideomotor learning revealed by fMRI. Neuroimage 46:290–298

Stefanatos, G. A. (2008). Speech Perceived through a Damaged Temporal Window: Lessons from Word Deafness and Aphasia. Semin Speech Lang. 29, 239-252.

Stein, D. & Fineberg, N. (2007). Obsessive-Compulsive Disorder, Oxford University Press.

Stepniewska, I. et al., (2011). Multiple Parietal–Frontal Pathways Mediate Grasping in Macaque Monkeys, J. Neuroscience, 31, 11660-11677.

Steriade, M. M. & McCarley, R. W. (2005). Brain Control of Wakefulness and Sleep, Springer.

Sunderland, A. et al. (2011). Tool use and action planning in apraxia.Neuropsychologia, 49, 1275-1286.

Sundram, F. et al. (2010). Neuroanatomical correlates of psychosis in temporal lobe epilepsy: voxel-based morphometry study. The British Journal of Psychiatry, 197: 482-492.

Sveller, C., et al., (2006). Relationship between language lateralization and handedness in left-hemispheric partial epilepsy, Neurology, 67, 1813-1817.

Takahashi, T., et al., (2009). Progressive gray matter reduction of the superior temporal gyrus during transition to psychosis, Arch Gen Psychiatry. 66(4):366-376.

Talmi D, Grady CL, Goshen-Gottstein Y, Moscovitch M. (2005). Neuroimaging the serial position curve. Psychol. Sci. 16:716–23.

Temple, E. et al. (2001). Disrupted neural responses to phonological and orthographic processing in dyslexic children: an fMRI study. NeuroReport, 12, 299–307.

Tesak, J., & Code, C. (2008). Milestones in the history of aphasia: Theories and protagonists. Psychology Press.

Tettamanti M., et al., (2009). Syntax without language: neurobiological evidence for cross-domain syntactic computations. Cortex. 2009 Jul-Aug;45(7):825-38.

Thomas, G. et al., (2008). Reduced structural connectivity in ventral visual cortex in congenital prosopagnosia. Nature Neuroscience, 12, 29 - 31

Tourette Syndrome Classification Study Group (1993). "Definitions and classification of tic disorders". Arch Neurol. 50, 1013–16.

Tranel, D. et al., (2009). Neuroanatomical correlates of the Benton Facial Recognition Test and Judgment of Line Orientation Test. Journal of Clinical and Experimental Neuropsychology, 31, 219-233.

Trimble, M. (2005). Psychiatrische Epileptologie, Thieme Georg Verlag

Tsakiris, M. (2010).My body in the brain: A neurocognitive model of body-ownership. Neuro-psychologia, 48, 703-712.

Tsapkini, K., et al., (2011). Orthography and Semantics in Broca's Aphasia: Evidence from Mor-phological Processing. Procedia - Social and Behavioral Sciences, 23, 55-56.

Tuzun, E. & Dalmau, J. (2007). Limbic Encephalitis and Variants: Classification, Diagnosis and Treatment, The Neurologist, 13: 261–271.

Tyler, L. & Marslen-Wilson, M. (2008) Fronto-temporal brain systems supporting spoken lan-guage comprehension. Phil. Trans. R. Soc. B, 363m 1037-1054.

Tyler, L. K., et al., (2010). Preserving syntactic processing across the adult life span: the modula-tion of the frontotemporal language system in the context of age-related atrophy.Cereb Cortex. 2010 Feb;20(2):352-64.

Tzourio-Mazoyer, N., et al., (2010). Left Hemisphere Lateralization for Language in Right-Hand-ers Is Controlled in Part by Familial Sinistrality, Manual Preference Strength, and Head Size. Journal of Neuroscience, 30, 13314-13318.

Uddin LQ, Kaplan JT, Molnar-Szakacs I, Zaidel E, Iacoboni M (2005) Self-face recognition activates a frontoparietal "mirror" network in the right hemisphere: an event-related fMRI study. Neuroimage 25: 926–35.

Ullman, M. T. (2007). Sex differences in the neurocognition of language, In Becker, J. B. et al. (eds). Sex Differences in the Brain: From Genes to Behavior . Oxford University Press.

Urgan, P. P. & Caplan, L. R. (2011). Brainstem Disorders, Springer.

Wallentin, M. (2009). Putative sex differences in verbal abilities and language cortex: A critical review. Brain and Language, 108, 175-183.

VanderHorst, V.G.J.M. et al. (2009). Estrogen receptor-[alpha] immunoreactive neurons in the brainstem and spinal cord of the female rhesus monkey: Species-specific characteristics, Neuro-science, 15, 798-810.

Van Der Mark, S. et al. (2011). The left occipitotemporal system in reading: Disruption of focal fMRI connectivity to left inferior frontal and inferior parietal language areas in children with dyslexia, Neuroimage, 54, 2426-2436.

van de Ven V, et al., (2009). Neural network of speech monitoring overlaps with overt speech production and comprehension networks: a sequential spatial and temporal ICA study.Neuro-image. 2009 Oct 1;47(4):1982-91

Van Essen, D. C. (2005). Corticocortical and thalamocortical information flow in the primate visual system. Progress in Brain Research, 149, 173-185.

van Veelen, N. M., et al., (2010). Left dorsolateral prefrontal cortex dysfunction in medication-naive schizophrenia.Schizophr Res. 123(1):22-29.

Verdon, V., et al., (2010). Neuroanatomy of hemispatial neglect and its functional components: a study using voxel-based lesion-symptom mapping. Brain, 133, 880-894.

Verstynen, T., et al. (2011). In Vivo Mapping of Microstructural Somatotopies in the Human Corticospinal Pathways. Journal of Neurophysiology, 105, 336-346

Vigneau, M., et al., (2010). What is right-hemisphere contribution to phonological, lexico-se-mantic, and sentence processing? Insights from a meta-analysis. NeuroImage.

Vocat, R. et al., (2010). Anosognosia for hemiplegia: a clinical-anatomical prospective study. Brain, 133, 3578-3579.

Von Kriegstein, K., Eger, E., Kleinschmidt, A., Giraud, A.L., (2003). Modulation of neural re-sponses to speech by directing attention to voices or verbal content. Brain Res. Cogn. Brain Res. 17, 48–55.

Wager TD, Smith EE. (2003). Neuroimaging studies of working memory: a meta-analysis. Neu-roimage 3:255–74.

Walkup JT, Mink JW, Hollenback PJ, (2006). Advances in Neurology, Vol. 99, Tourette Syn-drome. Lippincott, Williams & Wilkins. .

Wakabayashi, K., et al. (2009). Pathology of basal ganglia in neurodegenerative diseases, Brain

Nerve.61(4):429-39.

Waldie, K. E. & Hausmann, M. (2010) Right fronto-parietal dysfunction in children with ADHD and developmental dyslexia as determined by line bisection judgements,- Neuropsychologia, 48, 3650-3656.

Walker, R. H. (2010).The Differential Diagnosis of Chorea, Oxford University Press.

Wang, W. C. et al. (2010). The medial temporal lobe supports conceptual implicit memory Neuron, 68, 835-842.

Weiss-Blankenhorn, OP. H. & Fink G. R. (2008). Functional Imaging Insights into the Pathophysiology of Apraxia, Fortschritte der Neurologie, Psychiatrie , 76, 402-412.

Welter, M-L et al. (2011). Basal ganglia dysfunction in OCD: subthalamic neuronal activity correlates with symptoms severity and predicts high-frequency stimulation efficacy, Translational Psychiatry, 1, e5; doi:10.1038/tp.2011.5

Wheaton, L. et al., (2009). Left parietal activation related to planning, executing and suppressing praxis hand movements, Clinical Neurophysiology, 120, 980-986.

Wilkinson, D., et al., (2009). Unilateral damage to the right cerebral hemisphere disrupts the apprehension of whole faces and their component parts, Neuropsychologia, 47, 1701-1711.

Willems, R. M., et al., (2009). Differential roles for left inferior frontal and superior temporal cortex in multimodal integration of action and language. Neuroimage. 2009 Oct 1;47(4):1992-2004.

Whitney, C. et al., (2011). Heterogeneity of the Left Temporal Lobe in Semantic Representation and Control: Priming Multiple versus Single Meanings of Ambiguous Words, Cerebral Cortex, 21, 831-844.

Woloszyn, L., & Scheinberg, D. L. (2009) Neural dynamics in inferior temporal cortex during a visual working memory task. The Journal of Neuroscience, 29, 5494-5507

Yu, A. et al., (2011). Mapping anterior temporal lobe language areas with fMRI: A multicenter normative study. European J. Rad. 80, 441-444.

Zago, S. et al., (2010). A Cortically Blind Patient With Preserved Visual Imagery. Cognitive & Behavioral Neurologym 23, 44-48.

Zatorre, R.J. (2001). Neural specializations for tonal processing. Ann. NY Acad. Sci. 930, 193–210.

Zeman A. (2003). What is consciousness and what does it mean for the persistent vegetative state? Adv Clin Neurosci Rehabil, 3, 12-4

Zhang D, Zhang X, Sun X, Li Z, Wang Z, et al. (2004). Cross-modal temporal order memory for auditory digits and visual locations: an fMRI study. Hum. Brain Mapp. 22:280–89.

Zheng, Z.Z. et al., (2009). Functional overlap between regions involved in speech perception and in monitoring one's own voice during speech production, J Cogn Neurosci. 2010 Aug;22(8):1770-81.

CPSIA information can be obtained at www.ICGtesting.com
Printed in the USA
LVOW11s1510010913

350492LV00011B/525/P